LAMARCK'S SIGNATURE

FRONTIERS OF SCIENCE

Series editor: Paul Davies

Ripples on a Cosmic Sea: The Search for Gravitational Waves
by David Blair and Geoff McNamara

Patterns in the Sand:
Computers, Complexity, and Everyday Life
by Terry Bossomaier and David Green

Cosmic Bullets: High Energy Particles in Astrophysics
by Roger Clay and Bruce Dawson

The Feynman Processor:
Quantum Entanglement and the Computing Revolution
by Gerard J. Milburn

Lamarck's Signature: How Retrogenes Are Changing
Darwin's Natural Selection Paradigm
by Edward J. Steele, Robyn A. Lindley,
and Robert V. Blanden

Forthcoming:

Beginnings of Life: The Origins of Life on Earth and Mars
by Malcolm Walter

LAMARCK'S SIGNATURE

*How Retrogenes Are
Changing Darwin's Natural
Selection Paradigm*

EDWARD J. STEELE
ROBYN A. LINDLEY
ROBERT V. BLANDEN

HELIX BOOKS

PERSEUS BOOKS

Reading, Massachusetts

ISBN 0-7382-0014-X

Library of Congress Catalog Card Number: 98–87900

Published in Australia by Allen & Unwin Pty Ltd

Perseus Books is a member of the Perseus Books Group

Jacket design by Suzanne Heiser
Set in 10.5-point Plantin by DOCUPRO, Sydney

1 2 3 4 5 6 7 8 9–DOH–0201009998
First printing, September 1998

Find Helix Books on the World Wide Web at
http://www.aw.com/gb/

CONTENTS

Tables and figures vii
Preface xi
Acknowledgements xv
Dogma xvii

1 The twin legacies of Lamarck and Darwin 1
2 In the beginning there was RNA 25
3 Why the immune system is so interesting 58
4 The idea of 'clonal selection' 95
5 Somatic mutation 125
6 Soma-to-germline feedback 163
7 Beyond the immune system? 187

Epilogue 208
Appendix 223
Glossary 227
Notes 244
Bibliography 262
Index 272

TABLES AND FIGURES

TABLES

Dogma   xix

1.1   Outline of main phenomena and propositions in the structure of the traditional neo-Darwinian theory of evolution   4

1.2   Some basic genetic terms   17

2.1   The major discoveries and conceptual advances relevant to this book   26

3.1   Some basic immunological terms   61

5.1   Some basic terms relevant to somatic hypermutation and the soma-to-germline feedback loop   127

Appendix   The Genetic Code   223

FIGURES

1.1   The concept of Pangenesis   9

1.2   Schematic diagram showing the proposed 'Somatic Selection' soma-to-germline process in the immune system   10

vii

1.3	Weismann's Barrier	12
1.4	Theoretical outline of the evolutionary sequence for the increasing diversity of complex life forms on earth	20
2.1	Higher cells, bacterial cells and viruses	32
2.2	Enzymes (protein or RNA) are 'molecular machines' or catalysts	35
2.3	Diagram showing two representations of double-stranded DNA	38
2.4	The flow of genetic information—'Central Dogma' of molecular biology	43
2.5	DNA, RNA and proteins have three dimensions in space	48
2.6	Inheritance of mutations in DNA base sequences	50
2.7	Repair of damaged bases	54
3.1	Stylised representations of antibodies and killer T cells	65
3.2	Simple line diagram showing the protein structure of an antibody molecule	66
3.3	Basis of antibody specificity	71
3.4	The vertebrates in which studies have established the presence of mammalian-like immune systems	74
3.5	Diagram showing the protein structure of a pentameric IgM antibody molecule consisting of 10 HL heterodimers (5 monomeric IgM molecules)	76
3.6	Stick diagrams of antigen—antibody complexes	83

3.7 Diagram showing the binding of pentameric IgM to a multiple array of the same antigen   85

3.8 Schematic diagram of the time course of an antibody response   86

3.9 Human lymphoid system   90

3.10 Instructionalist model of antibody formation by Linus Pauling (1940)—the antigen template theory   93

4.1 Clonal Selection   97

4.2 Self-tolerance is not genetically determined   104

4.3 Peter Medawar's demonstration of acquired neonatal tolerance   106

4.4 A schematic outline of a bacterial gene and a 'single copy' gene in a higher cell   108

4.5 Schematic outline of the germline configuration and somatic rearrangement of V and C coding regions of antibody genes (heavy chain) in humans and mice   113

5.1 Diagram showing how point mutations can generate new mutant proteins with a different folding pattern   131

5.2 Diagram showing error rates of DNA and RNA synthesis   134

5.3 Wu-Kabat plots showing that hypervariable stretches in the variable regions of H and L chains of antibodies coincide with antigen-binding sites called 'complementarity-determining regions'   140

5.4 Somatic mutation and selection of mutant antibody-producing cells in a Germinal Centre 145

5.5 Diagram showing that the rearranged variable region V(D)J gene and adjacent flanking region DNA is the target of the mutator 151

5.6 Schematic diagram of the reverse transcriptase (RT)-mutatorsome 156

6.1 Paternal transmission of acquired immunological tolerance. Simplified protocol of the Gorczynski—Steele experiments 171

6.2 The DNA and amino acid variability (Wu-Kabat) profiles of germline chicken V-pseudogenes 176

6.3 Soma-to-germline feedback (for heavy chain) 178

7.1 Squatting postures of Oriental races and Australian Aborigines 194

7.2 Bone facets and squatting posture 195

Appendix

Translation of a messenger RNA base sequence into an amino acid sequence 224

PREFACE

The media-driven perception of 'Science' is that of a logical and precise activity which is constantly throwing up technological breakthroughs and products which we consume to improve our health and agriculture, to communicate, to better enjoy our leisure time, to travel, to improve our ability to fight war. At a time when the 'scientific impact' on our lives is all-pervasive—as 'Science' both solves and then creates new problems apparently *ad infinitum*—there is also widespread disenchantment particularly amongst young people. 'Science' apparently lacks those enduring humanist qualities which concern matters of the spirit or our emotions—'Science' cannot excite the mind and body as do the 'Arts', or 'Sport'. We disagree with these views because first, scientists being human have private and personal lives outside the laboratory and second, because the intellectual passions must be engaged in the pursuit of any decent scientific challenge. Science can be very emotive as it excites the life of the mind.

Let us rephrase: What is it that attracts us to a scientific life? What does it mean to be a scientist? Certainly one has to possess logical and analytical skills because the puzzles and problems created by the advance of scientific knowledge cannot be solved without these qualities. In our view 'Science' primarily is about highly refined curiosity. It is about asking pointed fundamental questions about the world. The mind *must* be engaged with a problem posed by the phenomena of the real world. Indeed the ability to ask the right questions can be considered as the 'art of science'. It is the quality and precision of the probing questions which often allow the correct solution to emerge. The euphoria of a new solution or discovery is often sufficient reward in itself.

Sometimes questions can in principle be answered but not so in practice as the technology (to do an experiment) is lacking; yet this does not negate the correctness of the questions and such an impasse may well stimulate further technological invention. But destructive critical analysis—the ability to tear open and expose falsehood or erroneous data—is essential and complements synthetic analysis. As scientists we must get to the heart of the matter and this often requires clearing away forests of falsehoods before we can see a new 'Truth'. Ideally, both the creative and destructive faculties need to reside within the same individual scientist. They amount to the fact that we are not afraid to confront the 'Truth' whatever that may be. This is a tall order because facing the 'Truth' is not the sort of thing we always enjoy, particularly if it means changing our mind.

We were approached by Ian Bowring of Allen & Unwin to write this book to communicate our ideas on evolution with the general public. Although we do not consider ourselves 'word-smiths' or 'semanticists' or even evolutionary 'game players' (when compared with the current genre of science book writers) we are nevertheless very interested in transmitting scientific reality as we perceive it in plain English. We have written the book for readers with no specialist knowledge. Yet we cannot deny that some sections (particularly Chapters 3, 4 and 5) are conceptually and intellectually challenging. In these sections we have devoted considerable space to explaining the scientific evidence and arguments necessary to fully understand Chapter 6 (the central message of the book). We include key concepts that we consider necessary to develop the general reader's understanding of molecular genetics and the immune system. If a section is 'heavy going' it could be postponed for later perusal, as the remainder of the book sets out a story more or less complete in itself.

At different times in this book we will pose questions which have excited us in our pursuit of a clearer understanding of how the antibody genes of the immune system have evolved. The antibody genes are a molecular metaphor of evolutionary change. Some of these questions (we think) have been answered; others will immediately spring to mind if the reader thinks our analyses or perceptions are faulty. If this book does stimulate such questions (or even alternative unexpected scenarios) then we will have considered this venture into the 'public domain' a success.

E.J. Steele, R.A. Lindley and R.V. Blanden
Wollongong and Canberra, January 1998

ACKNOWLEDGEMENTS

The intellectual threads and other influences which result in a book like this need acknowledgement. First, without the example set by Arthur Koestler a rational assault on Weismann's Barrier might never have been launched. Second, we are in debt to the seminal work of Melvin Cohn, Alistair Cunningham, Peter Bretscher and Rod Langman. We also wish to thank Jeff Pollard, Ann Wallace and Reg Gorczynski who helped start this saga twenty years ago; Gerry Both, for helping us with the molecular approach; Arno Mullbacher, who has been a tireless and creative devil's advocate; Ev Cochrane of *AEON* for his longstanding interest and for introducing us to the work of Frederic Wood Jones; Nicolas Rothwell for making us aware of the work of David Berlinski and other valuable suggestions; Frank Fenner for his help with the history of smallpox and additional background on Frederic Wood Jones; our students, Harry Rothenfluh and Paula Zylstra for their committed assistance and open-minded scholarship; Gordon Ada and Keith Tognetti for offering constructive suggestions on an early draft; Mary Halbmeyer of Natural Selection Editorial Services, who made many valuable editorial suggestions, most of which we used in the final draft; Emma Cotter and Karen Ward for editorial assistance; and our publisher Ian Bowring for encouraging us to communicate with the general reader. We thank all those publishers and trusts who gave us permission to reproduce previously published work; each of these is appropriately acknowledged in the legends and notes.

We acknowledge the following publishing houses for permission to quote from their books:

HarperCollins UK *(Collins Dictionary of Biology)*; Longman AddisonWesley, UK *(Henderson's Dictionary of Biological Terms)*; Penguin Books, UK *(The Penguin Dictionary of Biology)*; McGraw-Hill, NY *(McGraw-Hill Dictionary of Bioscience)*; Oxford University Press *(Oxford Dictionary of Biology)*; W.H. Freeman & Co for quotation (page 212) from Richard Dawkins' *The Extended Phenotype*; Kegan Paul International Ltd for quotation (page 188) from Frederic Wood Jones *Habit and Heritage*; Lippincott-Raven Publishers for quotation (page 70) from Bernard D. Davis, Renato Dulbecco, Herman N. Eisen, Harold S. Ginsberg, W. Barry Wood Jr *Microbiology*; Sinaur Associates Inc. for quotation (page 202) from WenHsiung Li and Dan Graur's *Fundamentals of Molecular Evolution*. Finally, we thank the Australian Research Council for financial support.

DOGMA

Without dogma our lives would be difficult if not impossible. A significant degree of dogmatic thinking is essential for social stability and harmony. In an open society, our dogmatic forms of behaviour underpin the very basis of all our nurturing institutions ranging from the rule of law and the provision of health care to the education of our children. The great religions of the world epitomise the durability of dogma in our lives; the life of the 'spirit' and the 'soul' is very important for large sections of mankind searching for meaning in a Universe apparently uncaring and without purpose.

In the world of ideas dogma focuses thinking. In a given domain it acts to eliminate or suppress alternative habits of thought. Successful scientific ideas often begin life as playful and potentially fruitful possibilities (circumventing the received teachings) because they appear to solve a pressing problem; they mature into acceptance and then freeze over, if you like, into dogma. Really useful dogmas allow us to economise our time and effort in not constantly reinventing the wheel. Scientific dogmas can be very useful and are only given up after vigorous and often sustained resistance. They are strongly defended—often literally to the death of the main intellectual players—precisely because they have allowed the fruitful development of a successful school of thought and have usually been of practical significance to mankind.

The practical and beneficial role of dogmatic thinking is often looked upon with disdain by innovative thinkers. However, the creative tension in all organic evolution and in the growth of scientific knowledge is between the need to preserve what has been proved and tested and the necessity, when the environment changes, to adjust body structures, physiologies, life strategies and theories to the new realities. The creative process in life and science, as put by Thomas Kuhn, is caused by 'the essential tension between tradition and change'.

We wrote this section after the book was completed, when we asked ourselves, 'Are we crazy or is the rest of the world crazy? Are we in fact actually seeing what we think we are seeing?' This is the standard refrain typical of all those involved in scientific discovery. The other muttering, often *sotto voce*, is, 'If it is obvious to us why hasn't the idea caught on like wildfire?' When one has been at the centre of a new concept for several decades it is easy to get caught up in a closed loop of complaint. That is why we must step back as we are doing here to celebrate the fact that without dogmas there would be no meaning to either our lives or our theories about the world.

The initial core ideas of this book emerged in the latter half of the 1970s. As the work has evolved through two decades, and as the data, the analyses and the rationales have strengthened, one often forgets that some of the guardian angels of the 'central dogmas', or, if you will, the keepers of the great reference books, have also been constructing and reconstructing their own conceptual fortresses—driven instinctively by the very important reasons we have just recounted above.

What do the dictionaries of biology say about the key ideas we will discuss in this book? After all, such dictionaries help provide a snapshot view of current dogmas.

These are often the sort of reference books all undergraduates and enlightened lay readers buy when they try to make sense of new knowledge or a new discipline they are entering.

It is appropriate, therefore, to start this book with a conceptual 'map' of the way ahead. In the following table we have summarised, often by direct quotation, what some of the world's dictionaries of biology have to say about such things as 'Darwinism', 'neo-Darwinism' and 'neo-Lamarckism'. They reflect the intellectual environment in which the present neo-Lamarckian ideas were conceived and which are just now coming to fruition after a long and difficult gestation. These ideas contradict a central pillar of neo-Darwinian thinking termed 'Weismann's Barrier' which forbids the inheritance of acquired characteristics (i.e. it forbids the flow of genetic information from somatic cells to germ cells). Eight categories of concept requiring institutional definition are listed. Some of you will be familiar with these terms already. The subtleties of these definitions should gradually spring to life as the message of this book unfolds. For example there appears to be a 'cultural' factor at work underpinning how scientific concepts in this area are fashioned. Thus Oxford, that bastion of conservative neo-Darwinism, makes no mention at all of Charles Darwin's Lamarckian theory of Pangenesis and generally refers to neo-Lamarckism in a disparaging manner. Yet, across the Atlantic, McGraw-Hill spell out the various terms in an 'objective' almost clinical style and in the process make no entries under either Weismanism or neo-Darwinism! Peruse the ideas in this table as you might look upon those geographical maps or family trees one often finds in the front pages of a novel. It is a hint of things to come.

	Darwinism	neo-Darwinism	neo-Lamarckism	Pangenesis
Oxford Dictionary of Biology 3rd Edition (1996)	Present-day species have evolved from simpler ancestral types by the process of natural selection acting on the variability found within population … the only unresolved problem was to explain how the variations in populations arose and were maintained from one generation to the next.	Current theory … formulated between 1920 and 1950 … makes use of modern knowledge of genes and chromosomes to explain the source of the genetic variation upon which selection works.	Any of the comparatively modern theories of evolution based on Lamarck's theory of the inheritance of acquired characteristics. These include the unfounded dogma of Lysenkoism and recent controversial experiments on the inheritance of acquired immunological tolerance in mice.	*No entry*
Penguin Dictionary of Biology 9th Edition (1994)	*Large historical entry—similar in its essentials to the Oxford Dictionary definition* [Natural variation pre-exists in the population and 'natural selection' allows the survival of the fittest parents to produce the next generation. Darwin's major problem was the origin of the genetic variation. This led to his proposing his theory of pangenesis.]	Current since early decades of 20th century, which combines Darwin's theory of evolution by natural selection with Mendelian heredity and post-Mendelian genetic theory. Accounts more successfully than Darwin was able for the origin and maintenance of variation within populations.	View, generally discredited, that acquired characters may be inherited. Notoriously espoused by the Stalinist biologist T.D. Lysenko … In another episode this century Austrian biologist P. Kammerer tried to demonstrate the phenomenon in midwife toads and sea squirts. More recently still, genetic transfer of acquired immunity in rats has been alleged. Experimental support for the view has generally been inconclusive, and sometimes even fabricated.	Theory adopted by Charles Darwin to provide for the genetic variation his theory of natural selection required. Basically Lamarckist, it supposed that every part of an organism produced 'gemmules' ('pangenes') which passed to the sex organs and, incorporated in the reproductive cells, were passed to the next generation. Modifications of the body, as through use and disuse, would result in appropriately modified gemmules being passed to offspring.
Collins Dictionary of Biology 2nd Edition (1995)	… different species of plants and animals have arisen by a process of slow and gradual changes over successive generations, brought about by Natural Selection. *A detailed entry follows similar to Oxford and Penguin definitions.*	A view of evolutionary theory that combines 'Mendelian Genetics' with 'Darwinism'. See 'Central Dogma'.	A modern attempt (for which there is scant evidence) to add genetical support to 'Lamarckism', that emphasises the supposed influence of environmental factors on genetic change.	*No entry*

	Darwinism	neo-Darwinism	neo-Lamarckism	Pangenesis
Henderson's Dictionary of Biological Terms 11th Edition (1995)	Based on (Charles Darwin's) theory that genetic variability exists within a species and the fact that organisms produce more offspring than can survive. Under any particular set of environmental pressures, those heritable characteristics favouring survival and successful reproduction would therefore be passed on to the next generation (natural selection).	The modern version of the Darwinian theory of evolution by natural selection, incorporating the principles of genetics and still placing emphasis on natural selection as a main driving force of evolution.	*No entry under this specific name. Entry under 'Lamarckism'.* The theory of evolution formulated by the French scientist J.B. de Lamarck in the 18th century, which embodied the principle, now known to be mistaken, that characteristics acquired by an organism during its lifetime can be inherited.	*Darwin's name not mentioned.* A now discarded theory that hereditary characteristics were carried and transmitted by gemmules from individual body cells.
McGraw-Hill Dictionary of Bioscience (1997) (extracted from 5th Edition *McGraw-Hill Dictionary of Scientific and Technical Terms*, 1994)	The theory of the origin and perpetuation of new species based on natural selection of those offspring best adapted to their environment because of genetic variation and consequent vigor.	*No entry*	*No entry, Entry under 'Lamarckism' reads:* The theory that organic evolution takes place through the inheritance of modifications caused by the environment, and the effects of use and disuse of organs.	Darwin's comprehensive theory of heredity and development, according to which all parts of the body give off gemmules which aggregate in the germ cells; during development, they are sorted out from one another and give rise to parts similar to those of their origin. 'Pangene': A hypothetical heredity-controlling protoplasmic particle proposed by Darwin.

	Central Dogma of molecular biology	Weismannism/ 'Weismann's Barrier'	Reverse transcriptase and generalised reverse transcription	Somatic mutation
Oxford Dictionary of Biology 3rd Edition (1996)	The basic belief originally held by molecular geneticists that flow of genetic information can only occur from DNA to RNA to proteins … now known, however, that information contained within RNA molecules of viruses can also flow back to DNA.	The theory of the *continuity of the germ plasm* … (which) proposes that the contents of reproductive cells (sperms, ova) are passed on unchanged from one generation to the next, unaffected by any changes undergone by the rest of the body. It thus rules out any possibility of the inheritance of acquired characteristics, and has become fundamental to neo-Darwinian theory.	An enzyme, occurring in retroviruses, that catalyses the formation of double-stranded DNA using the single RNA strand of the viral genome as template. This enables the viral genome to be inserted into the host's DNA and replicated by the host … used in genetic engineering for producing complementary DNA from messenger RNA.	'Somatic': Relating to all the cells of an animal or plant other than the reproductive cells. Thus a somatic mutation is one that is not heritable.
Penguin Dictionary of Biology 9th Edition (1994)	Proposal by F.H.C. Crick in 1958 that the flow of molecular information in biological systems is from DNA to RNA and then protein. RNA tumour viruses … since shown to transcribe single-stranded DNA from RNA templates by means of the enzyme reverse transcriptase, provide exceptions to the generalisation.	*Best entry under 'continuity of germ plasm'*: Heredity from one generation to another brought about by transference of a complex substance (germ plasm), itself part of the germ plasm of the original zygote which was not used in the construction of the animal's body, but was reserved unchanged for the formation of its germ cells.	*Entry in relation to reverse transcriptase in retroviruses similar to Oxford definition. General concept entered under Transposable elements*: Some eukaryotic transposable elements (retrotransposons or retro-posons), such as the Ty1 element in yeast transpose by reverse transcription of their RNA to yield a complementary DNA chain which inserts *(into genome)* after conversion to a double helix.	Body cell: any cell of multi-cellular organism other than gametes. Mutations in somatic cells do not generally play a significant role in evolution, being unlikely to be passed to further generations in gametes … asexual budding may produce new individuals with copies of somatic mutations and some plants may produce micro- and megaspores from somatic cells.

Note: We acknowledge the permission of each publishing house to reproduce, wholly or in part, these dictionary definitions.

	Central Dogma of molecular biology	Weismannism/ 'Weismann's Barrier'	Reverse transcriptase and generalised reverse transcription	Somatic mutation
Collins Dictionary of Biology 2nd Edition (1995)	The hypothesis (based on 'Weismannism') that genetical information flows only in one direction, from DNA to RNA to protein . . . in general, changes to protein structures produced by external forces are not inherited . . . The hypothesis has . . . been modified to account for activity of the enzyme 'Reverse Transcriptase' . . .	The theory, now considerably modified, which proposes that the germ cells are set apart at an early stage of development and are uninfluenced by characteristics acquired during life. The known action of chemicals and physical factors on chromosomes has resulted in modification of the theory into the 'Central Dogma'.	The process of synthesizing complementary DNA from an RNA template, by the enzyme reverse transcriptase . . . The process tends to be error-prone, because there is no 'editing' of the newly synthesized DNA and therefore mutations can accumulate. Reverse transcription is a reversal of the normal flow of genetic information from DNA to RNA. See 'Central Dogma'.	'Mutation' . . . In eucaryotes, if the alteration affects gametic cells the change is a genetic mutation and can be inherited; if body cells (nonsexual) are affected the mutation is called *somatic mutation* and will not normally be inherited.
Henderson's Dictionary of Biological Terms 11th Edition (1995)	The principle that the transfer of genetic information from DNA to RNA by transcription and from RNA to protein by translation is irreversible, now modified to take into account the transfer of information from RNA to DNA by reverse transcription carried out by some viruses.	The concepts of A. Weismann which deal chiefly with, the continuity of germ plasm and the non-transmissibility of acquired characteristics.	The synthesis of DNA on an RNA template, catalysed by the enzyme reverse transcriptase.	Entered under 'Mutation': Mutations occurring in body cells of multicellular organism are called somatic mutations and are only passed on to the immediate descendants of those cells; mutations occurring in germline cells can be inherited by the offspring.
McGraw-Hill Dictionary of Bioscience (1997)	The concept, subject to several exceptions, that genetic information is coded in self-replicating deoxyribonucleic acid and undergoes unidirectional transfer to messenger ribonucleic acids in transcription that act as templates for protein synthesis in translation.	*No entry*	The synthesis of deoxyribonucleic acid from a ribonucleic acid template.	'Somatic cell': Any cell of the body of an organism except the germ cells. 'Mutation': an abrupt change in the genotype *[read germline]* of an organism, not resulting from recombination; genetic material may undergo qualitative or quantitative alteration, or rearrangement.

CHAPTER 1

THE TWIN LEGACIES OF LAMARCK AND DARWIN

The past 45 years have seen an explosion in our knowledge of molecular genetics and this is transforming our view of the mechanism of inheritance and the evolution of life on Earth. The time is now ripe to analyse its revolutionary lessons for the general public. The scientific revolution initiated by Charles Darwin, that 'natural selection' is *the* driving force in evolution has now stagnated into dogma and requires updating with the new perspectives of the molecular revolution. This modification is coming particularly from the knowledge gained of the molecular genetics of the immune system. Indeed, as the twentieth century draws to a close, that other grandfather of evolutionary theory, the French biologist Jean Baptiste de Lamarck is, in one area of molecular biology at least, joining centre stage with Charles Darwin. Therefore it is now reasonable to entertain such previously heretical concepts and questions as: *Is there a Lamarckian principle at work in Nature? How permeable is Weismann's Barrier?* (the theoretical barrier between body cells and germ cells, sperm and eggs). *Can some acquired characteristics be inherited? If so, can we describe the process in molecular terms?*

Thirty years ago the great writer and philosopher Arthur Koestler posed these questions in a series of eloquent and profound books. Was he a lone voice crying in the intellectual wilderness, or are some of his scientific insights acquiring a new respectability?

Certainly within the body's 'internal universe' of cells and molecules of the immune system, neo-Lamarckian gene feedback concepts are now legitimate components of scientific explanation. Moreover, in this book we explain how aspects of the phenomenon known as 'somatic hypermutation' of antibody genes and the pro-posed soma-to-germline process are being articulated in the molecular genetic language of the copying of DNA into RNA and back again. We also explore the implica-tions of this new interpretative framework for organs and tissues beyond the immune system. Here it is premature to arrive at definitive conclusions, but clear propositions and questions can be posed relevant to the advance of knowledge in the future.

Charles Darwin himself, 130 years ago, made the first tentative steps towards a model of acquired inheri-tance. He called it 'Pangenesis', and it has a remarkable modern Lamarckist flavour. This little-known historical fact as well as the turbulent development and reception of Lamarck's ideas throughout the nineteenth and twen-tieth century is also recounted to remind the general reader that there is more to the ongoing debate on the mechanism of evolution than a slavish adherence to the current neo-Darwinian view (as instanced by the uncom-promising writings of Richard Dawkins and Daniel C. Dennett) that evolution proceeds *only* by the natural selection of *chance* events.

We will also attempt to strip away the jargon and explain to the general reader the basic concepts of molecular genetics that are rapidly changing our world view. Some definitions and biochemical terms are essen-tial and these are defined in the tables and figures as they are discussed. We urge the reader to refer to them and to the glossary if they are having difficulty with the text.

The key conceptual point considered is this: alterations in genes of *somatic* (body) cells of an animal appear to be transmitted to the genes of *germ cells* (eggs and sperm) and passed on genetically to offspring of future generations. This is essentially a molecular update of the original idea of the inheritance of acquired characteristics articulated clearly by Lamarck and later accepted by Darwin in his Pangenesis theory.

The main focus of the book will be the development of the scientific evidence which has resulted in this profound and fundamentally altered view of the way our genes behave. This is the new molecular genetics and it is based on our growing understanding of the complex molecular processes at the heart of the hereditary mechanism. However, we wish to emphasise here that so far our analysis and explanations are limited to the immune system of animals with backbones, the vertebrates.

★   ★   ★   ★

The traditional, and still largely correct view of the genetic mechanism of evolution, began with Charles Darwin's central assumption,[1] that the natural variation between and within species depends upon relatively stable genetic patterns. Then nature's 'blind reaper' acts to select the fittest parents to produce the next generation of offspring. This is the process of *natural selection*. For example, variations in body size, nutritional needs or the ability to evade predators are obvious characteristics among a myriad of other more subtle pre-existing variations which may enable the fittest to survive and pass on their characters to their offspring. In this way natural selection acts on pre-existing genetic variation in a population of organisms. The only genetic changes permitted in this conceptual scheme are conventionally

3

thought to be the result of chance—or random—mutations in the genes of the germ cells (eggs and sperm).
This is Charles Darwin's legacy, embodied in his *Origin
of Species* published in 1859 and his best-known contribution to our present day understanding of the
mechanism of evolution leading to higher plants and
animals. The modern neo-Darwinian view is summarised
in Table 1.1.

It would be fair to say that the natural selection

Table 1.1 Outline of main phenomena and propositions in the structure of the traditional neo-Darwinian theory of evolution

Inheritance The genetic material (DNA) can be passed unchanged from generation to generation.

Mutation Occasionally there is a permanent change in the inherited DNA message. Charles Darwin called mutations 'sports'. For example, an **A** in the DNA sequence could become a **G** resulting in a different genetic DNA sequence in that section of a chromosome, that is a change in the code in DNA and thus the RNA message, leading to change in amino acid sequence of protein and therefore a change in structure and/or function. Such mutations in the DNA base sequence provide an ongoing source of new genetic information on which evolutionary selection can act. These base changes are thought to be rare.

Random mixing of paternal and maternal chromosomes After fertilisation, a developing human embryo has 46 chromosomes. In the reproductive organs (the testes and ovaries) of sexually mature humans a special cell division process (meiosis) halves the number of chromosomes to 23 (called the haploid number), which are packaged into sperm in males and the eggs of females. When a sperm fertilises an egg, the number of chromosomes is restored to 46 (called the diploid number). The donation of chromosomes from each parent is a random selection that each parent had in turn inherited from their parents. In eggs the sex-determining chromosome is always an X. In sperm it may be either an X or a Y. The XX phenotype is female and the XY phenotype is male. All other characteristics are similarly determined by the chance inheritance of chromosomes. Thus the basis of the common observations we all make about our children hold true. That is, we might notice that, 'Tom seems to have his mother's face yet his father's hands!'

Recombination In sexually reproducing species physical exchanges (called 'crossovers' or 'recombinations') between the father's and the mother's chromosomes at the time the gametes (the sperms and eggs) are formed cause the shuffling of *pre-existing DNA sequences*. This process results in *new* combinations of inherited characteristics on each chromosome of an offspring.

Natural selection and evolution Not all genetic lines of organisms reproduce equally or at the same rate. Natural environmental conditions select those that have fitter pre-existing genetic lines and give them a *selective reproductive advantage*. There would be no 'direction' to cumulative hereditary changes if there was no selection. Evolution therefore is the result of the interaction of the genetic material with both the internal (cellular, within the body) and external environment in which the organism develops, and depends on selection of the fittest gene combination.

Isolation Free interbreeding in a large population, which enlarges the gene pool in which natural selection operates, is not possible in many genetic lines because of intrinsic and extrinsic isolation factors, the simplest being geographic separation of genetic lines from a common stock. For example, Darwin's various finch populations on the different Galapagos Islands became isolated, small, inbreeding populations. This was noted by Darwin to be essential for the emergence of new species.

Drift DNA sequences, whether they be large tracts of bases or single base differences, can be lost by accident. Genetic drift is part of the random background noise degrading the integrity of DNA sequences.

Weismann's Barrier is inviolate Acquired somatic modifications in a multicellular organism *cannot* be inherited.

Source: Adapted and modified from R.D. Alexander (1979) *Darwinism and Human Affairs*. University of Washington Press, Seattle.

world view has both guided and dominated much of our thinking in areas of genetics and evolution for most of this century. However in 1809, fifty years before the publication of Darwin's *Origin of Species*, the French biologist Jean Bapiste de Lamarck published the first coherent mechanistic outline of evolutionary change.[2] Indeed, Lamarck can be considered to be the father of the idea of species transformation. He did not conceive of a mechanism based on 'natural selection' but articulated very clearly a long-held view amongst many biologists and philosophers, that acquired characters can be transmitted to future generations of offspring. In modern parlance this means, for example, that changed body size or shape in a higher animal acquired as a consequence of changed feeding or nutritional habits, may be passed on to offspring.

We therefore introduce the two, not necessarily incompatible, concepts: the traditional 'neo-Darwinian' idea that evolutionary genetic variability pre-exists before the selective force acts (natural selection) *versus* the previously rejected 'Lamarckian' view of the generation of genetic variability at the same time as the selective force acts. The latter concept is particularly relevant to the immune system where it is now clear that the selective force/environmental stimulus (an infectious disease) exists at the same time as the appearance of new somatic genetic variability (mutated genes encoding antibodies against infection).

Thus, in historical terms, there has always been an alternative view on how evolutionary change in a population of organisms takes place. Such a view might explain why some species have been able to undergo an apparent rapid genetic transformation when sudden environmental changes, or catastrophes, have occurred in nature, or explain the rapid development of different breeds of domestic animal such as dogs, chickens, horses, cattle, sheep, goats, etc. In this book we will examine how feasible such changes are in terms of modern molecular genetics. This is not to say that Darwin's central and seminal idea of natural selection of random pre-existing variation is invalid. On the contrary, the Darwinian idea will be shown to be essential to the 'Lamarckian concept' of a gene feedback loop. Both Darwin and Lamarck were probably 'right'. Their twin legacies act to reinforce and complement each other in

6

the newly emerging view of evolutionary genetic change in the immune system.

It is a historical fact that Charles Darwin himself, and his grandfather before him, were Lamarckists. Darwin accepted Lamarck's theory of use and disuse of bodily parts coupled to a process of acquired inheritance. Lamarckian arguments of use and disuse appear in many places in his *Origin of Species* (1859). Ten years after its publication, he published his theory of Pangenesis.[3] This is a seminal contribution which is often expunged from the scientific literature by neo-Darwinists who seem to be deeply embarrassed that the originator of the theory of natural selection found it necessary to embrace a Lamarck-ian heresy to explain the origin of genetic variation. The spur for this theory arose from Darwin's acute observation of the exquisite variations and adaptations of plants and animals, particularly the latter, when domesticated by man. In Chapter 2 of *Origin of Species* in a section entitled 'Effects of Habit and of the Use or Disuse of Parts; Correlated Variation; Inheritance' he clearly outlines his view of the general way inheritance is effected:

> Changed habits produce an inherited effect as in the period of the flowering of plants when transported from one climate to another. With animals the increased use or disuse of parts has had a marked influence; thus I find in the domestic duck that the bones of the wing weigh less and the bones of the leg more, in proportion to the whole skeleton, than do the same bones in the wild-duck; and this change may be safely attributed to the domestic duck flying much less, and walking more, than its wild parents.

And he goes on to say:

> The great and inherited development of the udders in cows and goats in countries where they are habitually

7

milked, in comparison with these organs in other countries, is probably another instance of the effects of use.

In 1868 Darwin considered that during a somatic change necessary for a particular adaptation, the body cells of the target organ would be excited and emit genetic material in the form of what he called 'gemmules' (also called 'pangenes') or minute representations of each normal or altered bodily component. These were discharged by the active organ into the bloodstream and in the course of their circulation around the body they would enter the germ cells and be transmitted as part of the genetic endowment to the next generation. Figure 1.1 illustrates Darwin's concept of Pangenesis. As we will reveal in the following chapters, Darwin was prescient not only with respect to natural selection of random variants but also with his idea of the role of somatically emitted gemmules registering an inherited effect on what we would now call the 'germline DNA'.

More recently, a modern molecular view of the Pangenesis idea was articulated in the Somatic Selection hypothesis first proposed by Ted Steele in the late 1970s (Steele 1979). This hypothesis proposes a mechanism to explain the genetic evolution of antibody variable genes (V genes) via a soma-to-germline gene feedback loop. The mechanisms allow for the creation of new genetic variant animals in response to foreign microbial invaders from the external environment. That is, it provides a plausible hypothesis for Lamarckian inheritance based on contemporary molecular knowledge. Although Ted was initially rebuked by many members of the scientific establishment for his 'heretical ideas', the emerging data over the past twenty years has provided support for this hypothesis. There is now no doubt that the genes (made of DNA sequences) which encode the proteins for recognition of foreign invaders (antibodies) undergo

Figure 1.1 The concept of Pangenesis—Charles Darwin, 1868

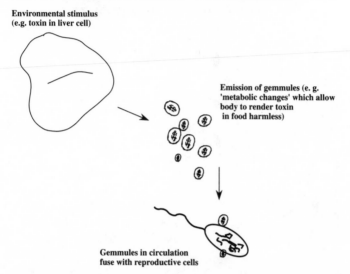

Environmental stimulus
(e.g. toxin in liver cell)

Emission of gemmules (e. g.
'metabolic changes' which allow
body to render toxin
in food harmless)

Gemmules in circulation
fuse with reproductive cells

The theory of 'Pangenesis' was published by Charles Darwin to provide an explanation for the origin of the biological variation on which natural selection acts. The basic idea of his 'use and disuse' theory involved increased activity of the target organ caused by some form of environmental stimulation (e.g. toxins in a changed diet). The changed metabolic activity of the target tissue (in this case the liver) would then lead to an adaptation resulting in the organ releasing 'gemmules' or 'pangenes' which would enter the circulation and fuse with germ cells. The changed function of the organ is thus incorporated on the germ cells to be inherited by the offspring. Darwin therefore provided a Lamarckian mechanism for the genetic transmission of acquired characteristics.

rapid somatic gene mutation as a result of being activated by the antigens of the invading infectious agent. The most recent data strongly suggest that antibody gene mutations are passed back to the germline DNA in a process involving 'reverse transcription'. Figure 1.2 illustrates the mechanisms that enable somatic changes to be copied and introduced into germline genes.

9

Figure 1.2 Schematic diagram showing the proposed 'Somatic Selection' soma-to-germline process in the immune system

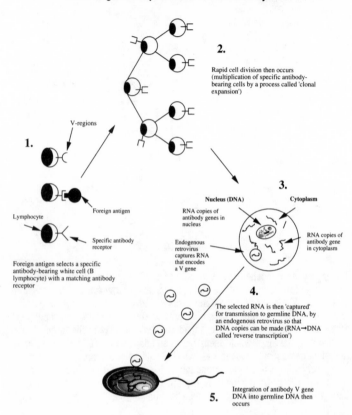

2.

Rapid cell division then occurs (multiplication of specific antibody-bearing cells by a process called 'clonal expansion')

V-regions

1.

3.

Nucleus (DNA)   Cytoplasm

RNA copies of antibody genes in nucleus

Endogenous retrovirus captures RNA that encodes a V gene

RNA copies of antibody gene in cytoplasm

Lymphocyte

Foreign antigen

Specific antibody receptor

Foreign antigen selects a specific antibody-bearing white cell (B lymphocyte) with a matching antibody receptor

4.

The selected RNA is then 'captured' for transmission to germline DNA, by an endogenous retrovirus so that DNA copies can be made (RNA→DNA called 'reverse transcription')

5. Integration of antibody V gene DNA into germline DNA then occurs

1. A diverse array of B lymphocytes exists prior to a foreign antigen entering the system. Each cell expresses on its surface membrane antibodies of one specificity. The variable (V)-region genes encode those parts of the antibody which form the antigen-combining site (as shown). A foreign antigen binds to those B cells which have complementary antibodies—these cells are therefore 'selected' in a Darwinian manner ('clonal selection').

2. After antigen binding the B cell is activated and divides producing progeny which in turn also divide producing more identical progeny cells. A clone of identical cells expressing that antibody is produced ('clonal expansion').

10

Individual cells of the clone can mutate their variable region genes (somatic mutations), which can themselves be selected by antigen to produce a new clone.

3. Within cells of the clone, RNA copies of antibody V genes are made in the nucleus. Mature (processed) versions of these messenger RNA molecules are exported to the cytoplasm where they are translated into sequences of amino acids making up the protein chains of the antibody (see Appendix for details of translation of RNA sequences into an amino acid sequence).

4. RNA molecules encoding V genes (nucleus or cytoplasm) may be 'captured' by harmless endogenous RNA retroviruses (produced by the cell) and complementary DNA copies of the RNA made by the viral enzyme reverse transcriptase (these are called retrotranscripts or cDNAs).

5. Copies of the antibody V genes are then transferred to sperm or egg cells by the virus and integrated into the germline DNA for transmission to progeny organisms.

(See Chapter 6 and Figure 6.3 for greater detail.)

Reverse transcription had a stormy reception from conservative scientists when it was first observed by Howard Temin in the late 1950s, but after a Nobel Prize was awarded to Temin and Baltimore (1975) it has become widely recognised as an essential process in the replication of retroviruses (such as HIV) and other cellular events. (Retroviruses are so called because genetic information flows from RNA to DNA, the reverse of the normal direction, DNA to RNA, in all living cells.) The main purpose of this book is to convey how the new molecular genetics is gradually eroding the edifices constructed in the past by neo-Darwinists to support the concept of natural selection acting on *random* genetic variations as the *only* mechanistic agent of evolutionary change. In the process, we are witnessing the birth of a new paradigm in evolutionary theory in the immune system based on a combination of Darwinian and Lamarckian concepts of evolution.

Given these developments, why then has the Lamarckian view been so controversial? Although it is not within the scope of our story to give a complete

Figure 1.3 Weismann's Barrier

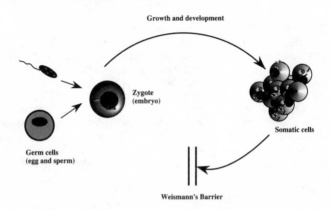

In 1885 August Weismann erected his famous tissue barrier protecting the germ cells from any form of somatic influence. In its modern articulation, this barrier is assumed to prevent somatic genes passing between somatic (body) cells and the germline (reproductive) cells.

historical description and analysis, several historical points can be noted. First, in 1885, three years after Darwin's death, German biologist August Weismann, responding to the challenge of Darwin's theory of Pangenesis, erected his now famous theoretical tissue barrier between somatic cells and germ cells (Figure 1.3). Weismann's Barrier was assumed to protect the germ cells from any type of change, physiological or genetic, within the body. The bulk of Weismann's experimental refutation focused on testing whether acquired parental mutilations could be inherited. For example, he cut off an appendage or similar tissue and showed that offspring did not inherit the mutilation. Of course any rational individual would have known that the Jewish custom of circumcision of young boys has never resulted in a baby boy being born without a foreskin. He also did other experiments to establish the continuity of the germ plasm on microscopic

12

animals such as *Hydra*. But it was his work on chopping off the tails of rats shortly after birth for which he is most famous. Weismann showed in breeding experiments extending over many generations, that such tail chopping at birth *never* produced tailless offspring. Critics of this experiment have pointed out that such experiments did not test Lamarck's idea. A short tail caused by chopping is a modification that was not produced by the rat. In contrast, Lamarck believed that only modifications produced by a *response* of the rat to the environment would be inherited. If Weismann had examined the behaviour (e.g. foraging, preening, movement) of the offspring of parental rats which had their tails removed he may well have observed subtle inherited modifications in behaviour.

The impermeability of Weismann's Barrier was allegedly strengthened in 1911 by the work of Castle and Phillips of Harvard University who transplanted ovaries from a black guinea pig into a white (albino) female after her ovaries had been removed.[4] The white foster mother, on mating to a normal white male produced several black offspring between six and twelve months following the operation. This is exactly the same result which would have occurred had the albino male been mated to a black female. Thus a new 'white' soma had no apparent effect on the germ cells in the transplanted 'black' ovaries. This type of acute experiment may be criticised on the same grounds as Weismann's tail chopping. There is no reason to suppose that an adaptive response in the grafted female would involve genetic control of coat colour.

The Lamarckian theme was next developed by the Austrian biologist Paul Kammerer with tragic consequences. He reported that altering the sexual and behavioural habits of marine and amphibious animals could result in progeny spontaneously showing the same altered behavioural or physical traits that their parents had acquired during their lifetime. His most famous experiment concerned the work on the midwife toad, *Alytes obstetricians*. Most toads and frogs mate in water, yet *Alytes* mates on dry land. Normal water-borne species of male toads mating in water firmly grasp around the waist of the female, maintaining the embrace for considerable time (days and weeks) until she spawns her eggs. To maintain this grip on the slippery female, the male develops calloused and horny-spined nuptial pads on its palms and fingers. However *Alytes*, mating on dry land, do not have or need these pads because the female's skin is dry and rough. Paul Kammerer reported that if *Alytes* were induced to copulate in water like other toads, over several generations they eventually developed nuptial pads as an acquired hereditary trait.

These experiments generated enormous envy and controversy in the early years of the twentieth century and finally culminated in accusations of scientific fraud by the geneticist Professor William Bateson of Cambridge University. According to Arthur Koestler, who documented this tragic story in his *Case of the Midwife Toad* (1971), these charges were never adequately proven, but the damage was done, ending in Kammerer's suicide a few years later (1926). More recently, this episode has been analysed in detail by Mark Gillman in his book *Envy as a Retarding Force in Science* (1996).

The third historical influence which has contributed to making Lamarckian thought a 'no-go' zone in science, can be attributed to the activities of the Russian plant breeder T.D. Lysenko. Lysenko was appointed by Joseph Stalin to improve Russian agriculture during the 1930s, 1940s and 1950s. Unfortunately, he applied questionable and non-scientific procedures in an effort to demonstrate crop yield improvement using a process referred to as 'vernalisation'. This involved the use of various seed treatments such as temperature and nutritional changes during germination, in an effort to produce a more productive plant which would then produce seed for better quality plants. The Lysenkian approach to reviving Lamarckian ideas completely alienated Western scientists and those few *bona fide* geneticists working within the Soviet Union. Lysenko compounded the problem by his notorious and ruthless pursuit of scientific opponents who ended up being executed by Stalin, or sent to the Gulag.[5]

These three epochal episodes in the history of science and society, we believe, could have acted in concert to inhibit the rational development of the original Lamarckian concepts embodying modern knowledge in the idea of a genetic feedback loop.

In parallel with these historical episodes, there emerge two underlying themes which have also influenced the scientific outcomes. The first concerns the iconic, dare we say exalted, position that Darwin himself has been elevated to by modern scientists, particularly in Britain. However, we also believe that much of this icon building was in fact necessary to embed the key concept of natural selection of random variations. Without this key concept, many biological phenomena—whether it be whole population effects, genetic sequence patterns or the behaviour of the immune system—would be difficult to explain. It was also seen by some as necessary to counter the naive creationist view that all species arose simultaneously and relatively recently.

The Darwinian view of evolution demands large time scales for the emergence of new forms and species, a fact consistent with the modern fossil and geological record. Indeed molecular evolutionary connections have been established between all extant living organisms. The generation of the molecular evidence which has made these conclusions possible has accelerated with the development of new and powerful computer-based gene sequencing machines introduced to laboratories throughout the world since the late 1980s. This technology has now made it possible for large amounts of accurate genetic (DNA) sequence information to be gathered by geneticists and molecular biologists in carefully controlled laboratory conditions. Much of this DNA sequence data has also been accumulated in large gene databases for wide reference on the Internet: 'Genbank' is one of these. The importance of the development of these new tools, and their role in revealing new scientific knowledge, was highlighted in 1993 when the Nobel Prize for Chemistry was awarded to Kary B. Mullis for the invention and development of the polymerase chain reaction (PCR). PCR techniques have been used in laboratories since the late 1980s and have made it possible to amplify the number of copies of a specified section of a DNA sequence of a chromosome by over a millionfold. The order of the sequence of nucleotide bases **A,G,C,** and **T** (the 'basic genetic letters' of a DNA sequence which, in groups of three, specify an amino acid) can then be rapidly determined with an automated DNA sequencer (see Table 1.2 and the glossary for definition of terms). So, PCR techniques now provide geneticists with a powerful new 'telescopic' tool to view the molecular structure and information content of the different sequence combinations of the nucleotide bases (**A,G,C,T**) which can loosely be called 'the genetic blueprint'. Indeed the PCR technique inspired the book and movie *Jurassic Park*, by showing the possibility (however fanciful at present) that preserved and very ancient DNA fossils could be amplified in large amounts allowing them to be cloned and pasted together to resurrect an extinct organism.

Over the last 25 years, many other new tools of molecular biology involving gene manipulation such as mutating, deleting, inserting, splicing and clonally amplifying pieces of DNA sequence, have also added

16

Table 1.2 Some basic genetic terms (also see Glossary)

DNA Deoxyribonucleic acid. A very long molecular polymer made up from the four informational 'bases' termed **A** (adenine), **G** (guanine), **C** (cytosine) and **T** (thymine). Chromosomes, which contain the genes, are made of long DNA sequences containing many millions of bases (in higher cells the chromosomal DNA is also complexed with protein). The DNA molecule in a chromosome is in the form of a double-stranded helix (see Chapter 2). A typical DNA base sequence might be:

5′-**AGCTTATTGCATAAGCGCGAT**-3′

5′ and 3′ These refer to the left-hand and right-hand ends respectively of a DNA or RNA base sequence.

The Genetic Code The portion of the chromosome which encodes the information specifying an amino acid sequence (a protein) is based on reading the base sequence as a series of sets of triplet bases or *codons* (see Appendix). For example the above sequence of triplets, reading left to right, would encode the amino acid sequence:

Ser-Leu-Leu-His-Lys-Arg-Asp

RNA Ribonucleic acid. A very long molecular polymer very similar to DNA made up from the informational bases, **A** (adenine), **G** (guanine), **C** (cytosine) and **U** (uracil). The 'messages' which encode amino acid (protein) sequences copied from the genes are made of RNA. RNA is usually single-stranded. For example, a typical messenger RNA base sequence might be:

5′-**AGCUUAUUGCAUAAGCGCGAU**-3′

Note: This is the same as the DNA sequence above except that **U** replaces **T**. **T** in DNA is functionally the same as **U** in an RNA sequence.

Nucleic acids The chemical term for all DNA and RNA molecules.

Proteins Proteins consist of long polymers of amino acids. Together with sugars and fats, proteins are the building blocks of normal body cells. An amino acid has a different chemical composition from nucleic acids.

Amino acids Amino acids are the basic building blocks that make up the amino acid polymer chains called proteins. The four base 'letters' in nucleic acids are read three-at-a-time as a series of triplet codons, each amino acid being encoded by one or more different triplets (see Genetic Code above and see Appendix). There are 20 standard amino acids making up the proteins in all living systems: **Gly** (glycine), **Ala** (alanine), **Val** (valine), **Leu** (leucine), **Ile** (isoleucine), **Pro** (proline), **Phe** (phenylalaine), **Tyr** (tyrosine), **Trp** (tryptophan), **Ser** (serine), **Thr** (threonine), **Cys** (cysteine), **Met** (methionine), **Asn** (asparagine),

---

Gln (glutamine), **Lys** (lysine), **Arg** (arginine), **His** (histidine), **Asp** (aspartate), and **Glu** (glutamate). A typical protein sequence (as shown above) might be:

Ser-Leu-Leu-His-Lys-Arg-Asp

Somatic cell Body cell of a multicellular organism.

Germ cell The reproductive cells (sperms, eggs) of a multicellular organism.

Antibody A protein produced by white blood cells (B lymphocytes) which helps rid the body of an infectious foreign invader.

---

greatly to our level of understanding of the processes that generate natural genetic variation. Indeed the physical manipulation of DNA sequences by molecular biologists is almost as straightforward as the manipulation of text, sentences and letters on the screen of a word processor. The coupling of new molecular genetic techniques with computer technology enables us to address new and important questions about our origins, about the pathology of viral diseases such as AIDS and about how genetic mutations arise. In the process, much information is emerging which sheds a new light on how we view evolutionary processes.

The DNA sequences for many similar types of genes (termed *homologous* genes) and the amino acid sequences of their protein products reveal connections from primitive bacteria through to higher plants and animals. For example, cytochrome c is a protein essential for the generation of intracellular energy (the energy molecule is called 'ATP') following the 'burning' of food molecules ingested by a living cell. By comparing the DNA (or amino acid) sequence of cytochrome c from many species it has been established that the cytochrome c gene (apart from minor genetic mutations) is essentially the same in *all* cells and species that generate ATP-based energy. This gene can be found in all forms of cellular life, from single free-living cells, to fungi, insects and

18

higher plants and animals, despite the fact that a billion or more years of evolutionary time may separate them from a common ancestor (see Figure 1.4). This pattern of molecular evolution, showing evidence for DNA sequence homology (relatedness), over time and across species type, is *repeated* for hundreds of well-studied genes, and now provides persuasive evidence for the fact of evolution. Amongst scientists, the passionate debate on evolution is not about its reality, but about the mechanisms at the molecular, cellular, organismal and population levels.

The second influence affecting the scientific acceptance of the central Lamarckian concept concerned the need for a cohesive framework for the establishment and development of modern genetics. Much of the earlier scientific work in genetics might not have been possible without the concept of a relatively stable 'gene'. This concept originated in 1859 from the work of the Augustinian monk Gregor Mendel, and was rediscovered at the turn of the century by the plant breeder Hugo de Vries among others, and was also enthusiastically embraced by the geneticist William Bateson. Over time, the notion of genes as stable Mendelian hereditary-determining units strung out on the chromosome 'like beads on a string' became widely accepted. Such genes, whilst expressed in the body of a mature plant or animal, were nevertheless considered to be protected in the germ cells of the reproductive tissue by Weismann's Barrier and were thought to be passed on *essentially unaltered* in each generation; apart from the normal mixing and scrambling of genes by genetic recombination between paternal and maternal chromosomes that occurs at the time the germ cells are being formed by a special cell division process, called 'meiosis' (see Table 1.1).

This idea of a stable gene facilitated the coherent development of modern genetics within the context of

Figure 1.4 Theoretical outline of the evolutionary sequence for the increasing diversity of complex life forms on earth

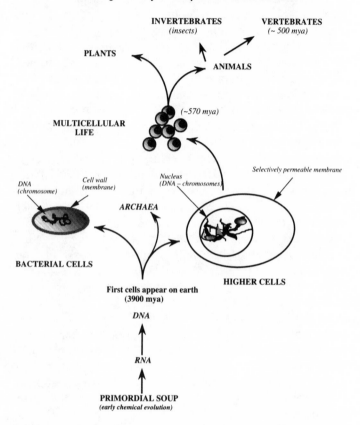

The most recent phylogenetic trees incorporating the main cell-based forms of life, the extremophiles or archaea, the true bacteria (prokaryotes) and higher cells with a membrane-bounded nucleus (eukaryotes) can be found in Woese (1994) and Madigan and Marrs (1997). Current estimates put the earliest cell-based form of life emerging at about 3.9 billion years ago (see Holland 1997).

Fred Hoyle and Chandra Wickramasinghe, in their recent book *Our Place in the Cosmos*, document the persuasive evidence that many bodies in the solar system, particularly comets, and also those planets with apparently organically-derived atmospheres (ammonia, methane), are teeming with cell-based life. Thus the phase of early chemical evolution may not have occurred on the Earth. Moreover the evidence presented by Hoyle and Wickramasinghe strongly suggests that the Earth has been infected on numerous occasions with cellular and possibly viral genetic systems in the past 4 billion years, particularly during passage of our solar system through massive molecular clouds in the Milky Way.

Darwin's ideas. The expression and determination of genetic relationships between parent and offspring would have been quite meaningless in a world where any type of somatic modification could be rapidly introduced into the germline. For example, in the case of Mendel's cross-breeding experiments between different varieties of pea plants, the new concepts of *recessive* and *dominant* genes for different characters would not have been observed if environmental influences were able to alter the genes at each generation.

It is against this brief historical backdrop that we will outline here our reasons for thinking that both Lamarckian concepts of environmentally triggered rapid genetic change, and soma-to-germline feedback, are necessary to rationalise the new molecular genetics as embodied in the vertebrate immune system. That is, we use our current molecular knowledge of the immune system to show that there is scientific evidence consistent with the Lamarckian notion of the inheritance of acquired characteristics.

However, the reader should also be aware these developments have been marred by ideological and dogmatic battles which began in the late 1970s. When Ted Steele reopened this contentious question in his 1979 book, *Somatic Selection and Adaptive Evolution,* the ideas generated a major international controversy played out in both the scientific and public arenas. He was reviled as 'The Bad Boy of Biology' by *Science Digest*, and as a 'Heretic' in *Science*. Yet others recognised the potential impact of such a theory. Colin Tudge of *New Scientist* wrote '. . . and if he [Steele] is right, we are begining a new chapter in biology'. The *Toronto Star* (4 February 1981) summed up the mood when it declared: 'He's [Steele] out to change the way evolution is viewed and at times it makes him as popular as a man standing in a crowd with a bomb.'

Scientists too have been divided in their views. However, Bob Blanden was one scientist who has clearly remained open-minded to the possibility that Lamarckian mechanisms may be at play—at least in the immune system. The collaborative partnership between Ted Steele and Bob Blanden has not only endured the turbulent scientific debates of the late 1970s and early 1980s, but it has also proved to be a fruitful intellectual partnership which has contributed enormously to the scientific data and arguments supporting the idea of Lamarckian inheritance in the immune system. As the data has accumulated, layer upon layer of the old world views has had to be discarded or fundamentally altered to accommodate the new data. Together with a team of collaborators spanning several disciplines, they remain galvanised in their pursuit of furthering our understanding of the underlying genetic mechanisms. They are now confident enough to state that 'if acquired inheritance in the immune system is not a real phenomenon then the only *ad hoc* alternative would be to invoke an intelligent gene manipulator, or "divine intervener", as playing a role in evolution'.

Here we wish to step aside from this past. Our focus will be on the objective analyses of the molecular and genetic data which has helped to develop our understanding of the distinctively Lamarckian mechanisms of the immune system.

We begin in Chapter 2 with an elementary description of genes, and molecular and cellular biology that is an updated version of the basic laws of the direction of flow of genetic information in living systems, the so-called 'central dogma'. Some of the main molecular characters featuring in this story (such as DNA, RNA, the nucleotide bases **A,G,C** and **T/U**, as well as the proteins) have been introduced in Table 1.2. In Chapter 2 we will also outline current views on early chemical evolution where it appears that the first informational molecule capable of Darwinian evolution was in fact RNA. However, paradoxically, the primary genetic material in all cells and many viruses is DNA. Why is this?

In Chapters 3, 4 and 5 we discuss the intense Darwinian 'survival of the fittest' processes that go on within certain populations of white blood cells (lymphocytes) when our immune system fights off infectious diseases. Indeed, current research into the molecules and cells of the mammalian immune system, particularly in the age of AIDS, occupies the research endeavours of thousands of molecular and cell biologists. We will outline the current views on these molecular mechanisms and emphasise the complementary roles of Darwinian selection and Lamarckian concepts in Chapter 5.

In Chapter 6 we begin by asking: Do the germline V genes (the genes which encode the variable regions of antibody molecules) benefit from somatic mutation mechanisms operating on V genes expressed in the mobile, circulating, antibody-producing lymphocytes? We show that the current molecular data strongly suggest the repeated penetration of Weismann's Barrier over evolutionary time; at least for the V genes of the immune system. The central informational molecule in this drama is not so much the relatively stable double-stranded DNA molecule (which stores the primary genetic information embodied in the sequence of **A,G,C** and **T** bases) but the relatively unstable intermediate molecule, RNA.

In Chapter 7, we examine the interesting hereditary phenomena associated with anatomical structures in humans and other animals. We look at the evidence that male rodents with drug-induced diabetes or other endocrine disorders pass on their altered trait to offspring at high frequency. In this chapter we also review evidence *pro* and *con* the possibility that a soma-to-germline feedback loop exists for other organs and tissues in the body. This is akin to asking if Darwin's idea of Pangenesis has general validity. We also highlight the challenges that are now before us to further advance the science that is the new molecular genetics.

Finally in the Epilogue, we step back from the science. Here, we carefully, though speculatively, consider some of the broader implications of Lamarck's legacy. For example we consider the implications for genetic engineering involving somatic gene therapy and the concept of 'genetic responsibility'. We also take the opportunity to reply to certain contemporary critics.

CHAPTER 2

IN THE BEGINNING
THERE WAS RNA

The purpose of this chapter is to give life to the new
concepts and molecules discovered in molecular genetics
since the 1950s. We will try to limit this to the bare
essentials in a field which has grown to nebulous complexity. This is not an easy task because unlike many of
the natural phenomena and life-forms in the world around
us described by the cosmologist, the palaeontologist, the
anthropologist and the field naturalist, the raw material of
modern molecular and cell biology are invisible; they have
been patiently teased out of their microscopic environment
by sophisticated experimentation using the tools of biochemistry and molecular biology. This makes the task of
'translating the story' to the general public somewhat
daunting. Unlike the cosmologist, palaeontologist, field
naturalist or student of animal behaviour, who have rich
historical traditions and sets of commonly known facts to
call on, we are compelled to meticulously explain exactly
what our raw material is before we can hope to bring
forth the challenging new concepts in genetics and evolution which are now emerging in the life sciences. We
have opted to do this by using images and diagrams to
build up concepts. There are therefore somewhat more
figures and diagrams in this book than one might find in
other contemporary science books written for the general
reader.

Our hope and aim is to communicate the excitement and full significance of the new molecular genetics. The science is now pointing to the operation of a Lamarckian soma-to-germline process (body cell to reproductive cell genetic flow) in the immune system of higher animals. We will retrace some of the main conceptual breakthroughs that have enabled scientists to rapidly build up their knowledge and understanding of the fundamental steps of genetic inheritance and information flow. Most if not all of our knowledge of molecular genetics has been discovered in the past 45 years. A chronology highlighting the most important conceptual advances and discoveries during this time is shown in Table 2.1.

Table 2.1 The major discoveries and conceptual advances relevant to this book

1952	A. Hershey and M. Chase show that hereditary material in bacterial viruses is in fact DNA and not protein.
1953	J. Watson and F. Crick discover the double helix structure of DNA. P. Medawar and colleagues establish the phenomenon of acquired immunological tolerance in newborn mice.
1957–59	'Clonal Selection Theory' of acquired immunity predicting 'Darwinian' antigen selection of specific antibody-producing cells (M. Burnet, D. Talmage, N. Jerne).
1957–59	The Watson and Crick mechanism of replication of DNA double helix proved and discovery of enzyme DNA polymerase (DNA→DNA copying; M. Meselson, F. Stahl, A. Kornberg).
1959	H. Temin predicts existence of 'reverse transcriptase' (RNA→DNA copying).
1961	Messenger RNA discovered. This is the intermediate molecule between the gene and protein. Copying of DNA→RNA by the enzyme RNA polymerase. (S. Brenner, F. Jacob, M. Meselson, B. Hall, S. Spiegelman).
1961–66	The genetic code deciphered. The base sequence of the mRNA molecule is read three at a time (codon) into a sequence of amino acid or protein (M. Nirenberg, H.G. Khorana, F. Crick, S. Brenner).

Table 2.1 (continued)

1965	Genetic model of antibody genes predicting the V→C rearrangement (W. Dreyer, J. Bennett).
1970	Self vs non-self discrimination rationalised in the two-signal model of immune induction (P. Bretscher, M. Cohn).
1970	Reverse transcriptase discovered in RNA tumour viruses ('retroviruses'), i.e. the copying of RNA→DNA (H. Temin, D. Baltimore).
1968–74	Somatic mutation of antibody genes rationalised by M. Cohn. 'Wu-Kabat' patterns established in antibody V-regions (T.T. Wu and E. Kabat). Antigen-driven somatic hypermutation of antibody variable genes predicted by A. Cunningham.
1974–77	Unique split-gene arrangement of antibody genes established and Dreyer-Bennett V→C model proved. First DNA sequence data confirming that antibody variable genes undergo somatic mutation (S. Tonegawa).
1977	Genes in higher cells contain 'introns' or non-coding intervening sequences (R. Roberts, P. Sharp).
1979	'Somatic Selection Theory' of acquired inheritance predicting soma-to-germline information flow for antibody variable genes in the immune system (E. Steele).
1981	Widespread proof that antigen-driven somatic *hyper*mutation occurs in the antibody genes of the immune system.
1982	Widespread proof for non-viral 'generalised' reverse transcription (RNA→DNA).
1982–83	Discovery of 'ribozymes' or RNA-based enzymes (T. Cech, S. Altman).
1985	Invention/discovery of the PCR technique (K. Mullis).
1987	Reverse transcriptase model proposed for antigen-driven somatic hypermutation of antibody variable genes (E. Steele, J. Pollard).
1987	Unifying general theory for the evolution of the immune system, the 'Protecton Concept' (R. Langman, M. Cohn).
1992–	Widespread emergence of evidence consistent with the reverse transcriptase model of somatic hypermutation. Molecular data accumulate supporting the 'Somatic Selection Theory' (H. Rothenfluh, R. Blanden, E. Steele).
1996–	The 'Weiller Algorithm' reveals the recombination signature in germline V-genes which supports the integration pattern predicted by the 'Somatic Selection Theory' (G. Weiller, R. Blanden, H. Rothenfluh, P. Zylstra, E. Steele).

Source: For a history of the period 1953–1970 see H.F. Judson *The Eighth Day of Creation*, Simon & Schuster, New York, 1979.

We must also make clear that we will be talking about complex information-handling processes going on inside living cells at the micron scale (a micron is one-millionth of a metre). The cells themselves can actually be seen under a standard light microscope (magnification about ×100 to ×600). If the DNA contained in the 46 human chromosomes were laid out end to end it would stretch several metres in length. Thus there is a very large amount of genetic detail embedded in these long DNA strands, yet they are folded and compacted into the tiny space of the cell nucleus, which is only a few microns in diameter. We will not dwell on the complex rules governing how the cell manipulates and replicates such enormous molecular polymers (that is, the chromosomes). Here it is sufficient to say that the process of chromosome replication and cell division called mitosis (to produce replica daughter cells) occurs in 5 to 20 hours depending on the cell type.

Thus, the 'genetic blueprints' defining the great variety of living forms observed are determined by complex living information systems that are contained in the cell. This system is analogous to the large amounts of computer code stored in small semiconductor chips which form a part of the central processing unit of a computer. Our understanding of the 'laws' of the genetic code and its translation into the biological products of living organisms has been revealed to scientists in parallel with the development of computer technologies that have exploded into our lives. However, nature has had a wondrously sophisticated storage and retrieval information system long before silicon chips were thought of. In fact, silicon chips are only now reaching a scale of 1 micron for mass production. Indeed, we are probably all aware of the talk of a new generation of computers and software every five years or so. Similar paradigm-shifting changes in thinking have (and are) taking place in modern molecular and cell biology. These epochal shifts have given rise to a new level of awareness of genetic function that is the focus of this chapter.

LIVING SYSTEMS CARRY OUT THEIR BUSINESS IN A CONTROLLED MEMBRANE-BOUNDED MICROENVIRONMENT

The first major step in understanding modern molecular genetics was the revelation of the complexity, structure and role of the single cell. At some point several billion years ago cell-based life began to flourish on earth. An integral part of living cells is the survival of the replicating RNA-based then DNA-based linear information molecules. Here we will refer to the totality of genetic information in the chromosomes of an organism as the 'genome'.

The selectively permeable membrane completely surrounding a cell provides a controlled environment, a solution of salts and other molecules in water, within which the genomes can be replicated and within which all the enzymes and multi-molecular machines (the 'organelles') can be synthesised and maintained. Food molecules (for example, carbohydrates and fats) are continuously imported by complex controlled transport processes into the cell, where they are used or stored. These constantly replenish pools of molecular building blocks, such as the bases **A**, **G**, **C**, **T** or **U**, (see Table 1.2). The oily water-insoluble cell membrane prevents the dispersion of the cell's contents into the wider environment.[1] It also supports a multitude of specialised surface molecules. They span the membrane, performing necessary functions to enable the cell to interface successfully with its external environment. The universe of molecules in the cell's environment can be large and diverse. Some cells may have tens if not

29

hundreds of different specific surface receptors that allow them to sense and respond to environmental stimuli. Some of these receptors also act as transport shuttles allowing the specific movement of particular molecules across the cell membrane.

There are a myriad of chemical reactions constantly occurring within the cell. Thousands of simultaneous reactions are exquisitely coordinated and choreographed in time and space like a complex computer program. Each reaction achieves a particular goal which contributes to the cell's growth and survival. If you were to visit a typical university department of biochemistry you would see large wall charts with criss-crossed interlinking arrows showing the direction of and relationship between many of the known chemical transformations that can occur within a cell.

What happens when we eat a bar of chocolate and assimilate the sugar? Biochemists have established that a typical cell extracts the useable energy from the glucose molecules by employing 30 or more specific enzyme-catalysed reactions. Each step is designed to strip the glucose of its energy, which is bound up in the chemical bonds binding the atoms of the molecule together. These specific chemical reactions are part of the process of 'molecular feeding' which has evolved to extract the maximum amount of useable energy.

Single bacterial cells are little different in their basic function from more specialised (differentiated) cells that make up the different parts and tissues of a multicellular organism. They are just smaller and contain a genome with fewer genes. Because there are fewer reactions to coordinate, the simpler bacterial cells also have the potential to grow and divide very rapidly in 20 to 30 minutes when the environment is favourable. This ability to divide rapidly makes them dangerous disease-causing agents if they invade a multicellular host organism. But this also makes them, and the tiny viruses which replicate within them (called 'phages'), the extraordinarily useful experimental tools that laid the foundations of modern molecular genetics.

Figure 2.1 shows 'viruses' in relation to bacterial cells and higher cells. A virus is simply a number of genes, embodied in a nucleic acid (either DNA or RNA) genome, depending upon the type of virus, wrapped up in a protein or membrane coat. By themselves, however, viruses do not grow and divide. All viruses are parasites; they need to enter the environment of a host cell to be able to replicate. This fact allowed Hershey and Chase to prove in 1952 that the genetic material of a virus is nucleic acid, not protein (see Table 2.1).[2] Some viruses infect bacterial cells and others infect the cells of higher multicellular organisms such as plants and animals. Some viruses, such as the influenza virus and the human immune deficiency virus (HIV) associated with AIDS can appear to be very clever in their capacity to rapidly mutate to evade the immune system.

When a virus enters a cell it uncoats its genome and proceeds to take over the cell's metabolic pathways for its own benefit. (A computer virus can achieve the same objectives by only revealing its active presence after it has entered a specific 'host' part of a computer.) The DNA or RNA of the virus is replicated in the cell and viral proteins are made. Mature progeny virus particles are assembled and are exported from the cell, where they go on to infect other cells. Some really dangerous viruses, like influenza, kill the cell as they replicate. Others coexist peacefully within the living cell they infect and continuously bud from the cell membrane as progeny virus. Still others, like herpes, may lie dormant (latent) within living cells for years before replicating and emerging. However, the ultimate viral parasites are the retroviruses. Some, like HIV, make a DNA copy of their RNA genome (by

31

Figure 2.1 Higher cells, bacterial cells and viruses

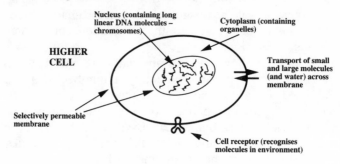

HIGHER CELL

Nucleus (containing long linear DNA molecules – chromosomes)

Cytoplasm (containing organelles)

Transport of small and large molecules (and water) across membrane

Selectively permeable membrane

Cell receptor (recognises molecules in environment)

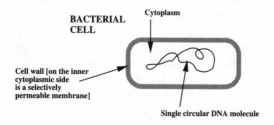

BACTERIAL CELL

Cytoplasm

Cell wall [on the inner cytoplasmic side is a selectively permeable membrane]

Single circular DNA molecule

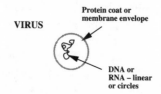

VIRUS

Protein coat or membrane envelope

DNA or RNA – linear or circles

Higher cells (also called eukaryotes) are 10–100 times larger than bacterial cells (termed prokaryotes). A selectively permeable membrane (composed of arrays of lipid [fat] molecules) containing protein receptors regulates the traffic of molecules, including water, into and out of the cell. A more rigid cell wall surrounds the bacterial cell membrane. Plant cells also have an outer rigid cell wall (not shown). Higher cells have a membrane-bounded nucleus containing the chromosomes (long linear DNA molecules). Bacterial cells (also called prokaryotes) have a single circular chromosome. In higher cells, RNA copies of genes are made in the nucleus (by the process of transcription) and transported to the cytoplasm after they have been processed (see Figures 4.4, 4.5). These messenger mRNA molecules are then translated into proteins on special molecular machines or 'organelles' called ribosomes in the cytoplasm (see Appendix). In bacterial cells RNA and proteins are all made in the cytoplasm as there is no membrane surrounding the single chromosome. Viruses are parasites which generally infect and take over cells using the cell's molecular constituents to reproduce more virus particles. They are very small, about 10–50 times smaller than the cells which they infect. They consist only of a protein coat or lipid membrane envelope and a genome which directs their replication (to produce progeny virus particles) only within a cellular environment. Viruses come in a variety of shapes and sizes. Their genome can be either DNA or RNA which can be double-stranded or single-stranded, circular or linear. Viruses which replicate in bacterial cells are given the special name 'bacteriophage'. Some viruses are harmless, such as the endogenous RNA retroviruses which are encoded in the genome of normal cells (as DNA) and are produced (as RNA transcripts) by normal cells, such as antigen-stimulated B lymphocytes (see Figure 1.2).

a process called 'reverse transcription', see page 11) and insert or integrate this DNA copy into the chromosome(s) in the cell nucleus where it is replicated as part of normal cell division.

There are three principles which are central to the process of molecular genetics:

- every single chemical reaction in biology is facilitated by a specific catalyst;
- DNA is composed of nucleotide bases which are structured in pairs; and
- genetic information flows from our genes outwards to proteins.

These central tenets form the basis of our understanding of how inheritance actually works.

EVERY SINGLE CHEMICAL REACTION IN BIOLOGY IS FACILITATED BY A SPECIFIC CATALYST

One of the most significant rules of molecular genetics has taken many years to confirm. It is now axiomatic that for every single chemical reaction that occurs in biology, there exists a specific protein enzyme (or enzyme-like molecules of RNA) to promote it. Such agents of change are called catalysts. This is a very important concept. It means that *no* multistep biological process occurs in nature by chance. These processes are regulated by a series of binding interactions between particular molecules which possess shaped faces that fit together like two pieces of a three-dimensional jigsaw puzzle.

Catalysts allow a chemical reaction to proceed more efficiently and quickly. Thus, in the absence of the enzyme catalyst, molecule A would not react immediately with molecule B to give the product AB. It might take place by chance but we might have to wait millions of years for the outcome if it occurred at all. However, in the presence of the specific enzyme, which has 'docking' sites for A and B (see Figure 2.2) the enzyme 'bends' and brings the two molecules so close together they have no alternative but to react, forming a chemical bond between them. Enzymes therefore are genetically encoded tools responsible for promoting the single reaction they have been selected to execute. In turn, a group of enzymes may be synchronised and coordinated to carry out a complex set of reactions that enable them to collectively perform much greater tasks, such as the extraction of energy from a chocolate bar when it is consumed. In other words, a single enzyme can be compared to a single instruction that forms a part of a larger computer program. However, there is one fundamental difference.

34

Figure 2.2 Enzymes (protein or RNA) are 'molecular machines' or catalysts

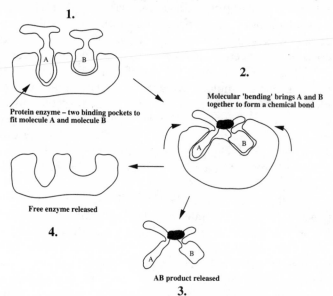

A protein enzyme is depicted bringing about a chemical bond between two different molecules, A and B, so as to produce the compound molecule AB. Some enzymes are made of RNA and they are called 'ribozymes'. **1.** The enzyme has two binding pockets to fit substrates A and B which dock with the enzyme. **2.** The enzyme facilitates the formation of a chemical bond between A and B by changing shape and creating the appropriate alignment of molecular interactions and forces. **3.** The product, AB, is released. **4.** The free enzyme can be re-utilised.

The creation of the computerised instruction set was conceived by a creative and goal-oriented human act. In the case of the enzymes, which are encoded by DNA or RNA sequences, their specific functionality has been selected and fine-tuned over millions (maybe even billions) of years of evolution, through the process of natural selection. A computer program is an act of intelligence dependent on human perception, while molecular processes which drive biological evolution occur through collisions between molecules or by 'binding in the dark'.

There are thousands of specific enzyme catalysts in the cell. For example, there are those that create building blocks for genes, those that allow food molecules to be recognised and taken up by the cell from the environment, and those that metabolise molecules inside the cell. Most biological catalysts are made of proteins, as are many of the structural components in a cell. Some catalysts are made of RNA.

In the late 1950s, Howard Temin predicted the process of 'reverse transcription' when he noticed that his RNA-containing tumour viruses simply disappeared when he added them to cells in tissue culture. He predicted that the virus copied its RNA genome into DNA and then integrated this DNA copy into the cells' chromosomes (the original virus RNA genome being eventually lost or degraded). He therefore predicted there must be an enzyme which copies the RNA sequence into a DNA base sequence. After a ten-year search, it was Howard Temin himself who discovered such an enzyme in the late 1960s. For this work, he was awarded the Nobel Prize in 1975, sharing it with David Baltimore, who confirmed the existence of the enzyme reverse transcriptase. However, this example has revealed more about genetic information flow than was originally thought. More recently (in the last ten to fifteen years), it has become clear that RNA informational molecules also have a catalytic role. They can cut and splice themselves ('edit') and they are potentially capable of self-replication.

It is now generally accepted among biological scientists that for every biological reaction that has been discovered, there is, and for all of those to be discovered in the future there will be, a specific catalyst to perform the task. Indeed we would go so far as to assert that for any hypothesised reaction we wish to predict there has to be some catalyst, usually a protein, sometimes an RNA molecule, which carries it out. All known data support this assertion.

THE SIMPLE BASE PAIRING RULE AT THE HEART OF THE HEREDITARY MECHANISM

Another important rule governing the hereditary mechanism is the base pairing rule for the building blocks of DNA: **A** always pairs with **T**; **G** always pairs with **C**.

In all living cells a gene is encoded in a DNA base sequence. The long linear strands of DNA in chromosomes are sequences of nucleotide bases symbolised by the letters **A**, **G**, **C** and **T** (see Table 1.2). The nucleotide sequence in a gene dictates the amino acid sequence of a protein. Let us explain.

The three-dimensional (3-D) structure of DNA was discovered in 1953 by James Watson and Francis Crick. This discovery represents an epochal event comparable to the significance of Darwin's achievement in the history of biology. The structure of DNA turned out to be remarkably simple. In hindsight, it was also entirely logical given its role as the carrier of the hereditary blueprint. DNA was found to consist of two complementary strands wrapped around each other to form the famous 'double helix' shape (Figure 2.3). The repeating structures of the polymer backbone were found to consist of a long chain of the nucleotide bases called **A** (adenine), **G** (guanine), **C** (cytosine) and **T** (thymine).

The key feature of the 3-D DNA structure is that the two strands run anti-parallel to each other and they are *base-paired*. This pairing of the bases is at the heart

Figure 2.3 Diagram showing two representations of double-stranded DNA

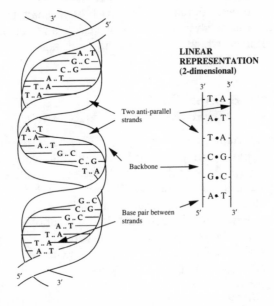

DOUBLE HELIX
(3-dimensional)

LINEAR REPRESENTATION
(2-dimensional)

Two anti-parallel strands

Backbone

Base pair between strands

The double helix on the left shows two base-paired, complementary strands. The strands in the right-handed helix are anti-parallel, one strand is 5′ to 3′, the other is 3′ to 5′. Notice that where there is an Adenosine base (**A**) on one strand, the base at the same position on the complementary strand is Thymine (**T**); where there is a Guanosine (**G**) on one strand there is always a Cytosine (**C**) at the same position of the complementary strand. The bases are part of a carbohydrate or ribose-containing nucleotide, each of which is linked together by 'phosphodiester bonds' forming the backbone of each strand. For convenience a molecule of DNA (always assumed to be double-stranded unless stated otherwise) is shown in the simplified 'Linear Representation' on the right. When the DNA is replicated the strands unwind revealing template sequences of free (unpaired) bases which are copied by DNA polymerase enzymes into complementary strands (Figure 2.4b).

38

of the genetic inheritance mechanism that makes possible the high stability and high fidelity copying of the genome. Double-stranded DNA is more stable and less easily degraded than single-stranded DNA or RNA. The base pairing rule states that wherever you have an **A** on one strand, there will be a **T** at the same relative position on the other strand; wherever you have a **G** on one strand you will have a **C** on the other strand. Notice that if the sequence of base letters on one strand is 5'-**AGCTAT**-3' then, on the other complementary strand, the sequence is 3'-**TCGATA**-5'. That is, they are anti-parallel and complementary to each other. The symbols for the ends of the strands, 5' and 3' (pronounced 'five prime' and 'three prime') have been defined in Table 1.2.

This is the extraordinarily simple law of DNA structure discovered by Watson and Crick for which they were awarded the Nobel Prize in Physiology or Medicine in 1962. However, the discovery of this pairing rule also depended very much on the empirical findings of Erwin Chargaff in the 1940s. He showed by chemical analysis that in all preparations of DNA purified from a number of different species, the number of molecules of **A** in a sample *always* equalled the number of molecules of **T**; and similarly, the numbers of **C** molecules *always* equalled the number of molecules of **G**. These were referred to at the time as 'Chargaff's rules'. Yet Chargaff failed to understand the significance of these observations for the structure of DNA, or the implications the observations had for the mechanism of genetic inheritance. This required the total 3-D framework of the double helix for these rules to make sense. There are many other instances in the history of science where important information does not make an impact at the time of its discovery, but lies 'dormant' until inspired creative insight reveals its true worth.

Thus the base pairing rules and anti-parallel arrangement of the complementary strands suggest how the DNA may be copied to form 'progeny' molecules with *identical sequences* (and thus information content) to the parent molecule. If you were to separate the strands, it can be imagined that the template sequence 3'-**TCGATA**-5' could act to direct the successive lining up of the bases **A**, then **G**, then **C**, then **T** and so on, to give the complementary sequence 5'-**AGCTAT**-3' (see Figure 2.3). Usually DNA is copied with extraordinary speed and fidelity when cells divide. However occasional mistakes, termed mutations, are made which alter the DNA sequence. These contribute to the natural genetic variation upon which natural selection acts.

DNA is therefore basically a linear information macromolecule much like the long strips of computer tape used in the first computers. The chromosome molecules are long double-stranded DNA polymers consisting of a sequence of millions of 'letters' (nucleotide bases). Such DNA sequences therefore form the coded instruction sets that are the basis of the genetic material for all known living cells on the face of the earth. The Human Genome Project—an international scientific effort being carried out in a number of specialised gene cloning and sequencing laboratories around the world—aims to completely specify the sequence of these letters in each of our 22 pairs of human chromosomes, plus the two sex chromosomes, X and Y. In computer databases all around the world and in the personal computer of the typical molecular biologist, hundreds of files of information contain thousands (maybe millions) of DNA sequence sets. Computer hardware and software advances now enable biologists to manipulate the sequences (for example, search for a particular sequence, snip it, or insert additional information). It is also fair to say that the evolution of modern computer technology has been essential for the recent succession of the many new discoveries in molecular biology (see Table 2.1).

THE CENTRAL DOGMA: GENETIC INFORMATION FLOWS FROM OUR GENES (NUCLEIC ACIDS) OUTWARDS TO PROTEINS

So far, we have described how our genome, our hereditary blueprint, is made of DNA. We have also said that most chemical reactions are mediated by specific protein enzymes. But how is the information in DNA converted into the production of the thousands of different proteins that are essential for the function and growth of living cells, and the development of myriads of different cell types that constitute the tissues of complex higher animals?

This is a crucial question that we could only begin to answer after the structure of DNA became known in 1953. By the early 1960s, Francis Crick, Sydney Brenner and their associates, together with H. Gobind Khorana and Marshall Nirenberg, worked out how information in the computer-tape-like DNA molecule could be 'translated' from a linear sequence of nucleotide letters (**GCTGGACTAATC**) into a corresponding sequence of amino acids (**Ala, Gly, Leu, Ileu**). This was a monumental deciphering task which reached its conclusion in 1966 when the genetic code rules were finally elucidated. The answer was that the order of the bases in a DNA strand constituted a triplet code. A particular set of three bases in a certain order specified a particular amino acid in a protein chain. So, in this example, **GCT** codes for **Ala** (alanine), **GGA** codes for **Gly** (glycine) and so on (see Table 1.2 and Appendix).

It was demonstrated by experiment that an *intermediary* information molecule between DNA and protein was involved. This intermediary turned out to be RNA.

41

However, unlike DNA, cellular RNA is made up of only single strands; yet, it is made up of the *same type* of basic building blocks as DNA, and its chemical composition was found to be very similar to DNA. It contains the bases **A** (adenine), **G** (guanine), **C** (cytosine) and **U** (uracil). The association between the two nucleic acids is as follows. If the sequence of nucleotide bases in DNA is 3'-**TCGAATA**-5', then the sequence of nucleotide bases in the RNA sequence (which is copied from the DNA) will be 5'-**AGCUUAU**-3' where **T** (thymine) now becomes **U** (uracil). We mentioned this rule in Chapter 1, page 17.

The important concept here is that a sequence of nucleotide bases in DNA can have a *complementary* sequence in RNA. The process of copying a DNA sequence into an RNA sequence is called 'transcription'. This takes place in the cell's nucleus. Notice in Figure 2.4a that the flow of genetic information is:

$$DNA \leftrightarrow RNA \rightarrow Protein$$

Note also that while an RNA sequence can be copied back into a DNA sequence ('reverse transcription'), a sequence of amino acids in a protein can *never* act as a copying template for the reverse flow of protein sequence information into RNA.[3]

But all of this still does not answer our initial question. How is the base sequence of DNA, now embodied as a sequence in RNA (termed messenger RNA or mRNA), translated into protein? Much of the scientific evidence answering this question was due to the work of M. Nirenberg and H.G. Khorana in the early 1960s. However, as more information became available, the process whereby a protein sequence is produced, from the triplet code in messenger RNA, turned out to be extraordinarily complex. The overall process (called

42

Figure 2.4 The flow of genetic information—'Central Dogma' of molecular biology

a. CENTRAL DOGMA — FLOW OF GENETIC INFORMATION IN LIVING SYSTEMS

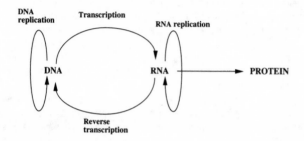

b. DNA REPLICATION

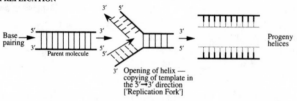

c. TRANSCRIPTION

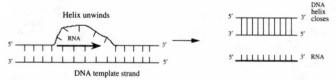

a. Depicted are the general rules of the direction of flow of genetic information, from nucleic acids (DNA/RNA) to proteins and *never* the reverse. Thus DNA base sequences and RNA base sequences can act as templates for the copying of other DNA or RNA base sequences, but amino acid sequences in proteins can *never* serve as a template for copying into an RNA (or DNA) base sequence. The main copying processes amongst nucleic acids are DNA→DNA

(DNA replication), DNA→RNA (transcription), RNA→RNA (RNA replication) and RNA→DNA (reverse transcription). Amino acid sequences making up proteins are specified by the sequences of bases in messenger mRNA molecules. This complex process, called 'translation', is carried out on ribosomes in the cytoplasm (see Appendix).

b. The first step in DNA replication involves the local unwinding of the double helix so as to expose the template sequences to be copied (this is called a 'replication fork'). A complex set of protein enzymes involving a DNA polymerase then copies each of the two DNA strands, synthesising a new complementary strand (about 1,000 bases are copied per second). Synthesis is always in the 5' to 3' direction as shown. When the end of the template molecule is reached the two progeny helices each contain one new strand and one parental strand.

c. Selected portions of the DNA sequence are copied into RNA either to make messenger RNA (mRNA) which encodes the information to specify a sequence of amino acids (see Appendix) or to make ribosomal or transfer RNA (rRNA and tRNA respectively) which are part of the molecular apparatus needed to translate mRNA into protein (see Appendix). The RNA polymerase copies the DNA template strand (there is local unwinding of the DNA helix to allow this to happen). Again the synthesis of the RNA is in the 5' to 3' direction so that the template strand of DNA which is copied is in the anti-parallel 3' to 5' orientation. A sequence of bases beyond the coding region of the gene specifies the termination of RNA synthesis.

'translation') is summarised in the Appendix. The simple rule is that the mRNA is exported from the nucleus into the cell cytoplasm where it is translated into a corresponding amino acid sequence (protein). The process is again similar to a computer tape being fed through a reading device which reads off the message three bases at a time. Each triplet of bases specifies a particular amino acid. In a cell the 'reading device' is a molecular organelle made of RNA and folded proteins called a 'ribosome'. A ribosome translates an mRNA into a protein sequence, a linear chain of amino acids. This chain must fold correctly into a particular three-dimensional stable structure to become a functional protein.

Also in the Appendix you will find the Genetic Code. It dictates which base triplet (codon) specifies a certain amino acid. The underlying Genetic Code, as far as it has been determined by experiment, is universal for all living organisms on earth from the smallest viruses and bacteria to plants and animals. During the evolution of life on earth, the code was selected as the optimum mechanism for information transfer leading to protein formation some 3 to 4 billion years ago and has never been changed or superseded. This provides a striking contrast with modern computer software, which is constantly being updated and superseded. It is interesting that Leslie Orgel and Francis Crick speculated about the origin of the code on earth in an article published about 25 years ago.[4] Given the ancient universality of the code they coined the term 'Directed Panspermia', speculating that the complex molecular machinery required to translate RNA into protein was seeded on earth by living organisms (bacteria) from elsewhere in the Universe—via either comets or space craft from a super-intelligent civilisation. A variation of this idea has also been championed by the astrophysicists Fred Hoyle and Chandra Wickramasinghe in their book *Life Cloud* and particularly in their more recent book *Our Place in the Cosmos* (see Figure 1.4).

Whatever the origin of the genetic code, it has become the basic code for all life on earth. It has led to the evolution of extraordinarily complex life-forms. Yet the basic rules have remained inviolate across species. It is also ironic that we have had to wait for the advent of the modern computer before research in molecular biology, based as it is on 'recombinant DNA technology' and gene cloning, could confirm the universality of the genetic code in all living species. Our own genetic mechanisms have enabled us to evolve to a level of conscious intelligence that allows us to conceive the idea of building up complex computer instruction sets based on preprogrammed rules using binary code. This has emerged in parallel with the development of our understanding of the rules of information flow governing molecular biology.

When we clone a human gene (say human insulin) we grow it up (clone it) in bacterial cells, thus making large amounts of the insulin-specific DNA. This indicates that the DNA replication machinery in bacteria treats a human DNA sequence as if it were a bacterial sequence and replicates it. Similarly, if we want to make large amounts of the insulin protein for treatment of diabetes, we now take the cloned human gene and 'express' it in bacteria. That is, we cause the bacteria to make the human insulin protein. In fact the bacteria make the *same* type of insulin with the *same* amino acid sequence. This confirms that the genetic code is read identically in bacterial and human cells, which are probably separated by 3600 million years or more of evolutionary divergence.

To summarise, the genes in the chromosomes are made of double-stranded DNA. As the cell grows and divides, copies are made of each strand of DNA and the new double-stranded helix is then passed on to each new daughter cell after cell division. As cells grow or change function to form different tissues in a multicellular organism they need to make thousands of different proteins which assemble into what we call 'multi-molecular machines'. Such protein (and RNA) assemblies coordinate and facilitate all of the chemical reactions that a cell needs to promote its survival, growth and development. Thus, genetic information encrypted as a linear sequence of bases in DNA specifies (via RNA) the complete universe of proteins that various cells need. Such informational molecules have generated the almost infinite variety of proteins—with different sequences, structures and functions—that have given rise to the evolution of the wondrous range and splendour of life-forms on earth.

While we have referred to linear sequences of DNA, RNA and protein, it must also be remembered that these informational and functional polymers have a three-dimensional structure in space. Thus double-stranded DNA is a right-handed helix (Figure 2.5a) and single-stranded RNA folds in space. RNA can form important functional structures by base pairing with nearby complementary sequences within its own length (see Figure 2.5b). The chain of amino acids in a protein also folds in space to give a distinctive three-dimensional shape (see Figure 2.5c) which, in the case of an enzyme, creates a specific molecular microenvironment for the particular chemical reactions it needs to catalyse. Therefore at the level of genetic information it is convenient for us to think in terms of a linear sequence, but at the level of the functional living process the game of life is played out in three-dimensional space.

Thus, in many ways, cells and multicellular organisms can be viewed as self-propagating and multifaceted information systems which can evolve over time. For example in our sequence of DNA on page 40, which was 5′-**AGCTAT**-3′, suppose the third base **C** mutates, to a **T**, then the sequence is different and all progeny sequences will be different. In other words, the mutant sequence can be propagated to all progeny molecules (see Figure 2.6). So at the molecular level, simple Darwinian selection of the fittest sequences can be imagined. Some of the most cogent proofs of the natural selection schema have been provided by mutant selection studies in molecular genetics.

REVERSE TRANSCRIPTION—THE COPYING OF RNA INTO DNA

In the decade after the discovery of the structure of DNA by Watson and Crick and the cracking of the

Figure 2.5 DNA, RNA and proteins have three dimensions in space

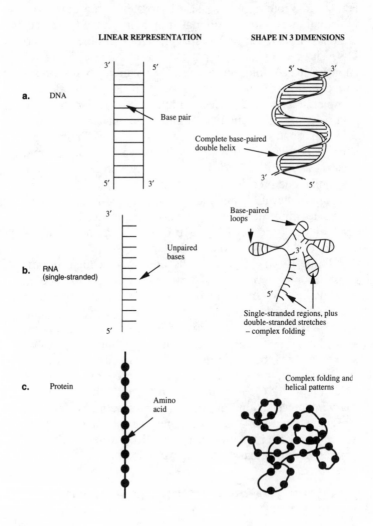

At the level of genetic and protein (amino acid) information, it is often convenient to think in terms of a linear sequence (those on the left); at the level of function within cells all these molecules have complex three-dimensional shapes.

a. Double-stranded DNA is a right-handed helix. However there are more complex orders of folding, particularly in higher cells, so that the chromosomes can be condensed and compacted within the confines of the nucleus. Depending on the stage of the division cycle of a cell, the helices are also complexed with a variety of different proteins, many of which determine which genes will be transcribed into mRNA.

b. Single-stranded RNA molecules may have a complex folding pattern involving local base-pairing of complementary sequences. These types of secondary structure are particularly important in protein synthesis in the functioning of the amino acid adaptor molecules, transfer RNA, and the RNA molecules involved in the structure and functioning of ribosomes (the rRNA molecules, see Appendix).

c. Sequences of amino acids in proteins adopt complex folding patterns in space, often involving local stretches of α-helix and other less rigid folding patterns termed 'β-pleated sheet'.

triplet code, the 'Central Dogma' of molecular biology (formulated by Watson in 1952) was validated. That is, genetic information only flows from nucleic acids (DNA or RNA) to protein—and *never* the reverse. This dictum, as we have said, holds for all biological systems.

These rules of genetic information flow are at the heart of molecular biology. The rules were only modified in an important and subtle way in 1970 with the confirmation of the existence of 'reverse transcription'. The discovery of reverse transcription was first made in tumour viruses isolated from mice and chickens. These RNA tumour viruses are now called 'retroviruses'. Their infectious cycle proceeds as follows. The virus enters the target cell, copies its RNA into DNA and then physically inserts the DNA copy of the virus into the DNA of the chromosome within the nucleus of the cell. Thus, when the cell divides, the integrated DNA copy of the viral genome is propagated and transmitted to daughter cells. In other words, the hereditary material of the virus is

Figure 2.6 Inheritance of mutations in DNA base sequences

a. Inheritance of a base error (point mutation)

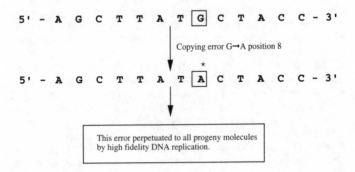

b. Inheritance of a four base deletion

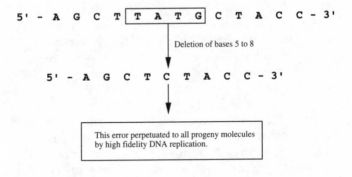

DNA replication is a high fidelity copying process, yet base substitutions and insertion/deletion events occasionally occur. Errors such as 'point mutations' (single base substitutions) are rare events, an incidence of about 1 base substitution for every 100 million to 100,000 million bases replicated (see Figure 5.2). The convention for presenting a DNA base sequence of a gene is to give the 5′ to 3′ strand (the strand which is in the same orientation as the mRNA—which itself does not act as the template strand for mRNA synthesis).

a. In the top example the 8th base, a **G**, has been substituted at that position by an **A**. If this base change occurs in the amino acid coding region of a gene it may lead to a different amino acid being specified in the protein following translation or it may lead to a premature termination of protein synthesis (that is, it may generate a translation stop codon, **TAA**, **TAG**, **TGA**, see Appendix).

b. A single base substitution is the simplest form of mutation; more complex mutations can involve the deletion or insertion of one or more bases. If insertions or deletions occur in the non-transcribed or 'flanking' regions of a gene they often have little genetic consequence; however if they occur in the region coding for amino acids the consequences are usually lethal because the protein sequence will now be quite different as insertion or deletion of bases always changes the triplet code reading-frame (see Appendix). Most often when the triplet reading-frame is changed (called a 'frame shift' mutation) this will cause the appearance further downstream of an inappropriate translation stop triplet (codon) such as **TAA**, **TAG** or **TGA** which will prematurely stop the synthesis of the protein (see Appendix).

permanently incorporated into the genome of the cells. Production of viral RNA by copying the integrated DNA can occur at a later time, leading to the production of a new infectious virus.

Why is this important? It is important because this is how the HIV virus becomes permanently integrated in the DNA of the nucleus of some normal cells of a human carrier. In this way HIV is propagated along with the cell's genome and is hidden from the immune response. This is partly why this virus is proving difficult to eradicate as each integrated DNA copy of the virus can later be used to produce *multiple* copies of HIV RNA by the normal process of transcription leading to the emergence of a new infectious virus from the hidden source. This is a viral 'Trojan horse' strategy. These viruses can then go on to infect and cripple other body cells. Other examples of retroviruses are those associated with breast cancer and leukaemia in human and other animals.

In late 1997, a publication in *Nature* from the group of Rolf Zinkernagel[5] in Zurich, Switzerland, showed that certain animal cells which contain the reverse trans-criptase enzyme of a retrovirus could produce DNA copies of the genome of another RNA virus which infected the cells. This intriguing discovery implies that reverse transcriptase enzymes encoded by retroviruses may potentially make DNA copies of other RNA mol-ecules present in the cell containing the enzyme. Thus, free DNA copies (called cDNA or retrotranscripts) of cellular genes potentially could be produced from mRNA templates. This idea is the core of the soma-to-germline theory of evolution (first proposed by Ted Steele in 1979). What was then a theory has now become one step closer to reality. We will describe the other molecular evidence supporting the theory in more detail in later chapters.

THE RNA WORLD—DOWNLOADING RNA GENETIC BLUEPRINTS TO DNA

Given our growing understanding of these processes, there is another question that continues to be asked by biologists. What came first: DNA, RNA or protein? There is a whole field of study devoted to this question. Here we concentrate on the main points that are of particular relevance to the new molecular genetics.

In the early 1980s Thomas Cech, Sydney Altman and their collaborators discovered enzymic properties associated with RNA molecules, which they called 'ribozymes'.[6] Since that time, there has been much elaboration of the concept of the 'RNA world'. In this view of life, the first living molecules were polymers of RNA. Proteins as we now know them, and DNA-containing chromosomes, did not exist. In the RNA world, the microscopic living forms were self-replicating RNA molecules of varying lengths, consisting of hundreds or thousands of bases. Replication of the linear RNA strand was mediated by a folded version of the *same* molecule (ribozyme). Thus, the first molecules capable of self-replication and thus Darwinian evolution were not the DNA double helices as we understand them today. Replication of such DNA helices requires a complex set of specific protein enzymes making up the 'DNA replication machine'.

The features being discovered of the hypothetical RNA world are therefore allowing molecular biologists to address an issue similar to the familiar question: Which came first, the chicken or the egg? Indeed, whatever the lucky accident was that produced an RNA polymer, this event immediately encapsulated both the chicken and the egg, as the molecule could perpetuate itself.

The discovery of the enzymic and self-replicating potentialities of RNA has completed the cycle of 30 to 40 years of speculation towards developing a rational understanding of the first steps involved in the molecular evolution of life. Thus, the discovery that single-stranded molecules of RNA would replicate themselves and mutate (accidentally change their base sequence) was functionally significant. It has now become evident that all nucleic acid copying processes involving RNA intermediates (transcription, RNA replication and reverse transcription, see Figure 2.4 a and c) are *error-prone*.

This is important here because all genes in cellular organisms are double-stranded DNA molecules which are replicated faithfully by a complex replication machine (consisting of about 30 or 40 different proteins). This high fidelity copying of DNA is achieved because a number of the DNA replicating-proteins deal exclusively with *editing and correcting* the newly produced copy of

53

Figure 2.7 Repair of damaged bases

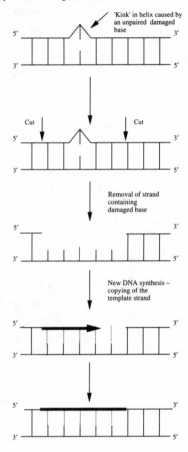

Errors in DNA base sequences which are propagated to progeny obviously contribute to genetic variation in the population of organisms and thus to evolution by natural selection. The errors or mutations are rare because the DNA replication apparatus has several sequence editors and checkers monitoring the molecular integrity of the double helix. The figure shows (schematically) one type of error-correcting process involving the removal of a chemically altered or damaged base which can no longer pair with its counterpart on the other strand. A short region around the unpaired (or damaged) base is removed by special cutting enzymes (called endonucleases) and the gap is filled in the 5' to 3' direction by repair DNA synthesis using the other strand as a copying template.

DNA. For example in DNA replication, if an **A** was inserted during copying instead of a **C**, the 'editor' enzyme can detect this and reinsert the correct base, because the normal base pairing process has been disrupted. Further, if chemical damage to a base occurs after the newly synthesised strand has base paired to the parent (template) strand a 'kink' is caused in the double helix prompting the editing functions of the enzyme complex to snip out the damaged sequence and resynthesise a good copy (see Figure 2.7). It is akin to the process that occurs to check data integrity during electronic message transmission. This is why gene mutations occurring solely at the level of DNA are extremely rare events. The mutations we see are those that have slipped through all of the normal editing and checking gates of the DNA replication machine. DNA replication is a very high fidelity copying process: in fact, for every 100 million to 1000 million base sequences replicated, only about one base error goes undetected.

Thus the picture we now have of the early RNA world is one almost of 'evolving chaos'—the survival of the fittest self-replicating molecule. Manfred Eigen and colleagues have done some illuminating research that has demonstrated how error-prone copying in RNA and environmental selection for function might give rise to a quasi-optimal population of RNA molecules. This population of RNA molecules rapidly evolves in a Darwinian manner to the changing environment.[7]

In comparison with double-stranded DNA, single-stranded RNA is chemically much more unstable. It is clear that at some point deep in the evolutionary past it became advantageous in an early life form to stabilise the hereditary blueprint. This was the seed of the transition from an RNA to a DNA world. An essential first step in the transition to a DNA world is the downloading of RNA sequence information to DNA sequence information, that is, the emergence of 'reverse transcription'. These early reverse transcriptases were most likely RNA enzymes (ribozymes). Later would come the emergence of the more complex protein-based reverse transcriptases.

However, for some RNA viruses such as influenza and the retroviruses the high error rates inherent in RNA production, together with other genomic strategies, provide a selective advantage. The high mutation rates (a base error is generated for every 1000 or so bases replicated) enable evasion of the immune system of infected vertebrate hosts. This has led to the retention of RNA as the viral genome, long after the evolution of protein enzymes. Very recently it has been shown that the enzymes which build the DNA sequences on the ends of vertebrate chromosomes (telomeres) by reverse transcription from an RNA template (discovered in the late 1980s by Elizabeth Blackburn) actually have a clear evolutionary relationship with viral enzymes which copy viral RNA sequences. This is another extraordinary example of the evolutionary significance of RNA.

Why have we put such an enormous emphasis on this single step in molecular evolution? We have done so because later in this book we reveal how reverse transcription is central to all modern views on Lamarckian genetic feedback enabling the inheritance of acquired somatic mutations. From a more philosophical perspective, it has also shown how the evolution of higher order

56

biological systems has required the step-by-step development of inbuilt genetic 'intelligence', or genetic
'consciousness',[8] which has ultimately given rise to
human consciousness and creativity.

CHAPTER 3

WHY THE IMMUNE SYSTEM IS SO INTERESTING

Our bodies are exposed to a relentless barrage of potentially invasive pathogenic viruses and bacteria. How does our immune system fight the constant war against such a multitude of pathogens? Moreover, in the course of maintaining a healthy body, the immune system is often far from being at harmony with itself. There is always the potential for full-scale 'immunological warfare', involving a battle within our own cells and tissues, as well as vigorous responses to fight off foreign invaders. How do our bodies quickly make whole armies of new antibodies to fight previously unknown pathogens? Those of us in good health remain so because of the complex and decisive role our immune system plays in maintaining the integrity of our body.

Such is the complexity and extraordinary 'intelligence' of the immune system of the vertebrates (animals with backbones) it enables pathogen recognition and regulation of immune responses, as well as generation and maintenance of immunological memory. The role of the immune system is to manage the war within. The fascination with the details of how this occurs is one of the reasons many Nobel prizes have been awarded for work that has furthered our understanding of these fundamental questions.

The battle starts during late development of the embryo. The cells of the immune system first encounter their own molecular components ('self antigens'). Shortly after birth and environmental exposure, the immune system must then adapt to combat environmental pathogens ('foreign antigens'). This situation, however, presents a fundamental problem for the immune system. Given that our immune cell populations (T and B lymphocytes) are 'designed' to fight a vast range of foreign pathogens, how then do they stop short of destroying 'self'? How do they discriminate between 'self' and 'non-self'? This is a fundamental problem that immunologists have been trying to solve for most of this century.

Another key question has also become evident. Given that the molecular forms of all pathogens cannot be predicted in advance, how does an animal respond to the unexpected? How does our body 'remember' the characteristics of previous pathogens so that they can be fought more effectively at the second—or subsequent—encounters? If you pause to think about these questions for a minute, then the answers to them have intrinsic scientific and philosophical relevance way beyond the narrow confines of the immune system. These are just some of the reasons why the immune system is so interesting to study.

In this and the next two chapters, we outline how our understanding of this specialised biological topic has developed, how the immune system of vertebrates has evolved and how modern molecular genetics is only now revealing some of the underlying principles and forces at play. As we have begun to answer each of these questions, our eyes have been opened anew to additional unsolved scientific, practical and ethical problems. It is also true that in the light of this new knowledge and data, many things we believed in the past about the immune system are now standing on shaky ground. Indeed the science of immunology is a rich biological metaphor for our contemporary understanding of genetics, development and evolution. Of particular relevance to this book, we explain how recent scientific discoveries are now challenging fundamental neo-Darwinian assumptions about evolution. In Chapter 6 we will go on to outline how the complex molecular apparatus of the immune system may permit acquired somatic genetic information to enter the germline and be inherited by future generations.

To begin this journey of explanation, we must introduce the reader to some important concepts. We need to discuss the main cells, protein molecules and genes that make up the immune system, the essential players in this fascinating game. We also define some of the basic terms used (see Tables 3.1 and 5.1). In Chapters 5 and 6, we will continue this intellectual journey and discuss in detail our most recent concepts explaining how somatic gene mutation takes place in antibody variable genes and how soma-to-germline gene feedback of such mutant DNA sequences may be achieved. These concepts form the basis for the central theme of the book—that characteristics acquired during one's lifetime may be passed on to offspring.

A BRIEF HISTORY OF VARIOLATION AND VACCINATION

In the eighteenth and nineteenth centuries, the concept of vaccination against infectious disease was given a scientific foundation. Deliberate smallpox inoculation as a protective measure against epidemic smallpox was first introduced to Europe by Lady Mary Wortley Montagu, an outspoken and quite unconventional intellectual diarist who herself had a severe attack of smallpox when 22 years old. This resulted in her losing her eyelashes and becoming severely pockmarked. In 1717 Lady Mary

Table 3.1 Some basic immunological terms (also see Glossary)

Antibody A protein produced by *B lymphocytes* (white blood cells) in response to foreign antigens (bacterial cells, virus particles and their toxin products). A typical structure is represented by immunoglobulin G *(IgG)*, a protein heterodimer—see Figure 3.2. There are many different antibodies, each of which recognises a different antigen.

Antigen/'self antigen' An antigen is a foreign protein or carbohydrate polymer—not a constituent of the 'self'—which stimulates the production of specific antibodies which bind specifically to the antigen and not other unrelated antigens. A 'self antigen' is a protein or carbohydrate which is part of the body. Immune responses against self antigens do not normally occur.

B lymphocyte Bone marrow-derived white blood cell which matures into antibody-producing cells. B cells can be bound and 'selected' by the antigen, leading to activation into antibody secretion or cell proliferation.

Clonal Selection/specificity of antibodies As B cells mature from the bone marrow they express surface antibody (Ig) molecules all of the same specificity in any individual cell, i.e. the *same* sets of variable region genes (see below) are used in each Ig molecule. A foreign antigen entering the system selects, in a 'Darwinian' manner, only those B cells to which it can bind. These cells are activated and divide (or proliferate) forming a *clone* of specific B cells, where all the progeny cells are *identical* in antibody specificity—see Figure 4.1.

Constant (C-) regions/immunoglobulin classes Amino acid sequences of an antibody molecule which do not take part in antigen binding and are therefore not subject to variation from one molecule to the next. The constant (C) regions of the heavy chain determine the function of the antibody, e.g. the ability to promote phagocytosis. C-regions of heavy chains can be classified into five general types of sequence (and therefore function) termed *'immunoglobulin class'* (see below).

Cytokines/interferons Protein molecules secreted usually by activated helper T cells which deliver activation and differentiation signals to other white blood cells including other T or B cells.

Haptens Small chemical groupings (benzene rings and their derivatives) prepared by chemists which by themselves are not able to induce antibodies. When physically bonded to larger 'carriers' (foreign proteins), the hapten can be used to induce anti-hapten antibodies.

Table 3.1 (continued)

Housekeeping gene   Refers to the large range of all *necessary* genes whose protein/RNA products are *essential* for the basic functions of the cell (for example the enzymes of the energy-generating pathways, the enzymes needed to replicate the DNA, etc.).

Immunoglobulin   Formal term for an antibody protein abbreviated as *Ig*. There are five 'classes' of immunoglobulin termed *IgM, IgG, IgD, IgA* and *IgE*. They are classified by differences in amino acid sequence in the C-regions of their H-chains; each Ig class has a different function (see Glossary).

Junk DNA/'flanking' region sequences   Those parts of the chromosomal DNA sequence which do not appear to code for proteins or other types of functional RNA (ribosomal RNA, transfer RNA). Another term to describe the DNA sequences that surround coding regions is *'flanking'* sequences.

Phagocyte/phagocytosis   Phagocytes are white blood cells which engulf (phagocytose) and digest bacteria and other cellular debris.

Protecton   A unit of antibody-mediated 'protection' defined as the minimum critical concentration of specific antibody in the blood required to protect against infection.

T lymphocyte   Thymus-derived white blood cell which expresses antibody-like surface molecules called *T cell receptors* (or TCRs). T cells come in two basic types, *killer T cells*, or *helper T cells*.

Transplantation antigens/killer T cells   Also called *'major histocompatibility antigens'* or *MHC antigens*. Expressed on most cells of the body. *Killer T cells* from one individual can recognise the MHC antigens of a non-identical individual and react to them, thus rejecting transplanted tissue. The MHC antigens also act as *'antigen presenting structures'*, e.g. presenting fragments of the viral proteins (peptides) to killer T cells, activating them and causing the killer T cells to then attack and lyse the virus-infected cell.

Variable (V-) region/antigen-combining site   Both H and L chains each contribute 'variable' amino acid sequences at their tips (see Figure 3.2). Also termed an *antigen-binding site*. V-regions form the antigen-combining site of an antibody (or T cell receptor).

travelled to Turkey with her husband, the British ambassador in Constantinople. She became fascinated by her discovery of the local custom called 'ingrafting'. Two weeks after arriving, in a famous letter dated 1 April 1717, she wrote to a friend:

> The small-pox, so fatal and so general amongst us, is here quite harmless . . . Old women . . . make it their business to perform the operation every autumn . . . people send to one another to know if any of their family has a mind to have the small-pox: . . . the old woman comes with a nut-shell full of the . . . small-pox and asks what veins you please to have opened . . . and in eight days' time they are as well as before (and immune from smallpox).[1]

Lady Mary realised the implications immediately (she called it 'variolation') and wrote to some doctors in England. She had both her son and daughter variolated by the physician Charles Maitland and she tried to encourage wider interest in the procedure in England. Despite initial opposition, she did manage to get royal permission for a trial involving six convicts from Newgate Prison, which was also carried out by Charles Maitland. This trial and another were successful and the Princess of Wales then arranged for the variolation of both her daughters. However, variolation had been practised in China and India for centuries prior to its introduction to Europe. Variolation involved taking the scabs of the sores of a smallpox sufferer and scratching the scab remains into the skin of a healthy uninfected person, who would usually suffer a non-lethal infection, recover and then be immune to further smallpox infections.

The great contribution made by Edward Jenner in 1798 was to show that smallpox could be prevented if a person was previously inoculated with cowpox—a milder, related virus. His studies were founded on the common observation that milkmaids who had recovered from cowpox never developed smallpox during epidemics. Jenner himself remembered being variolated as a child and becoming very ill, thus having first-hand motivation to improve matters. Jenner also practised variolation for many years before his breakthrough with cowpox vaccination, which is a far safer and more pleasant procedure. Since then, smallpox and polio vaccinations have been the success stories of preventative medicine. By 1977 smallpox had been eliminated by an international vaccination program organised by the World Health Organisation (WHO).[2]

The scientific basis for vaccination was firmly established by the second half of the nineteenth century by the father of modern microbiology and immunology, Louis Pasteur. In controlled trials he established the effectiveness of using 'weak' non-life-threatening forms of the disease-causing organisms (which he called 'attenuated' strains) to immunise against the virulent form of the disease and in 1881 proposed general use of the term 'vaccination' in honour of Jenner.[3] Attenuated bacterial or viral strains retain the same antigens, or molecular characteristics recognised by the immune system, as the original disease-causing strain but have lost the capacity to cause serious disease.

To the present day, some of the most successful vaccines are attenuated strains (e.g. vaccines against yellow fever and poliomyelitis). However, molecular genetic technology is now being used to insert DNA sequences encoding the antigens of interest into attentuated bacterial or viral vaccines or to produce pure protein antigens for use as vaccines. Some of these new generation vaccines are in clinical use. We now know that vaccination stimulates the inoculated individual to produce activated T lymphocytes and protective antibodies which help our bodies eliminate the foreign viral or bacterial infections (Figure 3.1). A sketch of the protein structure of a typical antibody is shown in Figure 3.2.

Figure 3.1 Stylised representations of antibodies and killer T cells

a.

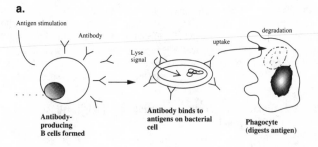

b.

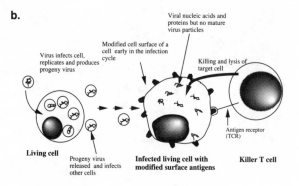

Highly simplified outlines of the main protective roles of antibodies and killer T cells (see Table 3.1 for basic terms).

a. Following antigen stimulation B cells are selected and activated to produce antigen-specific antibodies. In the case of bacterial cells acting as antigens (as shown) the antibodies bind to the surface antigens of the bacteria triggering two types of result. On the one hand the antibodies may co-opt other proteins in serum (called 'complement') which act with the antibodies to directly lyse and thus kill the bacterial cells. Alternatively, the bound antibodies may trigger engulfment and intracellular killing/digestion by special white cells called 'phagocytes' (the ingestion process being called 'phagocytosis').

b. Killer T cells protect against viral infections. The virus infects the cell and replicates. Virus proteins are synthesised. Progeny viruses are eventually assembled and released to infect neighbouring cells thus causing acute disease. Infected cells express cell surface transplantation antigens modified by forming complexes with viral protein fragments (termed 'peptides'). It is this modified cell surface antigen which is recognised by killer T cells through their cell surface T cell receptor (TCR), resulting in proliferation and activation of the T cells. When the activated killer T cell binds the target cell through its TCRs the target cell is lysed and killed. Since the modified surface antigen can be expressed well before mature virus particles are assembled, killer T cells can destroy the 'cell factory', prevent progeny virus being made and thus clear the infection.

Figure 3.2 Simple line diagram showing the protein structure of an antibody molecule

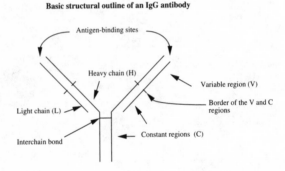

Basic structural outline of an IgG antibody

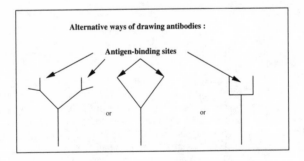

Antibodies are heterodimers because each basic subunit consists of two different protein chains, a heavy (H) and a light (L) chain (see Table 3.1). Two

H + L pairs make up the basic molecule often held together by an 'interchain bond' as shown. This structure is typical of IgG antibodies and cell surface bound forms of monomeric IgM (and IgD). Note that the antigen-binding site is made by the interaction of the variable regions of an H chain and an L chain. The constant regions of the H chain determine the immunoglobulin class and protective function of the antibody (see Table 3.1). The boxed figures at the bottom are simplified cartoons of an antibody often used in diagrams (see Figures 1.2, 3.1, 3.6, 3.7 and 4.1).

BLOOD TRANSFUSIONS AND THE RISE OF MODERN IMMUNOLOGY

By the beginning of the twentieth century it was evident to pioneers such as Paul Ehrlich and Elie Metchnikoff, that antibodies against cholera, a violent diarrhoea-causing disease of the small intestine, were highly specific, binding to the antigens of the cholera organisms but not other antigens. They shared the Nobel Prize for their discoveries of the specific nature of antibodies and the role of antibodies and phagocytic (scavenger) cells in immunity. The specificity of antibodies was comprehensively demonstrated in the early years of this century by Karl Landsteiner (1930 Nobel Prize), who also made a number of other important immunological discoveries. His most famous discovery concerned the system of human ABO antigens expressed on red blood cells which formed the rational basis of all cross-matching of donors and recipients for blood transfusions:

Antigens on red cells	'Naturally occurring' antibodies in the same individual's blood
A	anti-B
B	anti-A
AB	None
O (no A or B)	anti-A and anti-B

If A red blood cells are mixed with a blood serum containing anti-A antibodies, the red cells visibly clump together ('agglutinate'). However, A cells exposed to anti-B serum do not agglutinate. Agglutination could go on in the bloodstream if an A individual inadvertently were transfused with B blood, which would contain anti-A antibodies (with potentially severe, acute reactions). Ideally if we are typed as A we should receive A blood and likewise for B. In an emergency an A or B individual could be transfused with O red blood cells washed free of blood plasma (the fluid which contains the naturally occurring anti-A and anti-B antibodies). In O individuals the haemoglobin-carrying red cells do not have A or B blood group antigens on their surface, which if present would stimulate life-threatening anti-A and anti-B antibodies. Thus O is the universal donor type for red blood cells and, conversely, AB is the universal recipient.

Landsteiner's discovery helped to explain why so many transfusions were not successful. However, blood transfusion decisions are now more sophisticated and include matching for the Rh antigen which is involved in haemolytic disease of the newborn of an Rh+ve father and an Rh−ve mother. Indeed, since these early immunological experiences, modern laboratory research has taken us to a point where we have now isolated and fully characterised many of the protein molecules, cells and genes which contribute to the vertebrate immune system. We now know how antibodies and T cell receptors are produced as well as the DNA sequence structure of the genes involved. But several key questions underpinned the development of immunology, the first was: 'Just how many specific antibodies are there?'

HOW LARGE IS THE REPERTOIRE OF ANTIBODIES?

This important question was also raised by the work of Landsteiner. These experiments are less well known than his definition of the ABO system, yet they have had far-reaching consequences. Landsteiner conducted experiments showing the enormous range of specific antibodies that could be made by laboratory animals. Being skilled in organic chemistry, he chemically attached small man-made molecules of complex carbon rings, such as derivatives of benzene, to different types of protein antigens and demonstrated the production of specific antibodies to them in laboratory rabbits and mice. These small molecules (which he called 'haptens') by themselves usually did not evoke antibody production. Yet, when physically coupled to foreign protein carriers, the resulting hapten–protein conjugates became powerful antigens stimulating the production of copious quantities of antibodies, specific both for the hapten portion and the much larger protein.

By itself Landsteiner's production of antibodies against these antigens was not so remarkable, as immunologists had been producing antibodies by experimental vaccination for many years in rats, mice, rabbits, guinea pigs, goats and horses. The really striking finding was that he could elicit antibodies against all of the new chemicals and drugs being produced during those years by the fledgling pharmaceutical industry. These new antigens had never existed in nature! This result posed another scientific—if not philosophical—dilemma. Why should chemicals which never existed on the face of the earth until their production in the laboratories of the modern pharmaceutical chemist provoke an antibody response? There would not be any evolutionary selection pressure for antibody production against a substance that did not exist! Yet Landsteiner was able to show that laboratory animals could make antibodies to each chemical, binding specifically to each unique molecular structure.

The exquisite specificity of Landsteiner's antibodies can be demonstrated by simple laboratory experiments. For example, it was found that hapten **A** antibodies could not be removed from the antiserum by reacting it with haptens **B**, **C**, **D**, **E**, **F**, **G**, **H** . . . etc. That is, hapten **A** antibodies were specific and had little or no affinity with the range of other haptens considered. However, in more sensitive test-tube precipitation reactions, some cross reactivity patterns could be seen (see Figure 3.3). In this example a negative sign (–) indicates that no visible sediment appears in the test-tube; a positive sign or a series of them (+ to +++) refers to the visible intensity of the precipitate and indicates the presence of insoluble antigen-antibody complexes; a trace precipitate would be scored ±. Antigens **D** and **G** appear to be related to **A**, and indeed Landsteiner could show that **D** and **G** often shared a common molecular structure with **A**.

Landsteiner called these experiments 'Studies in Antigen Specificity' and, collectively, they have provided a basis for the first estimation of the size of the potential antibody repertoire. Thirty years ago, a standard immunology text of the day commented on the genetic and evolutionary implications of Landsteiner's results:

> It is difficult to see, however, what advantages accrue from the preservation of genes for the synthesis of antibodies to many recently synthesised organic chemicals—such as p-amino-benzoate, 2,4-dinitrobenzene, etc.—which have no obvious resemblance to microbial pathogens. The preservation of so much excess genetic baggage through eons

70

Figure 3.3 Basis of antibody specificity

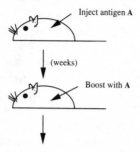

Test reactivity of blood serum by mixing with other antigens :

Antigen	Formation of an antibody-antigen precipitate
A	+++
B	-
C	-
D	+
E	-
F	-
G	±
H	-

Immunisation with a particular antigen (**A**) stimulates the production of antibodies that bind to **A** and not **B**, **C** etc. The simplest test for antibodies is to mix the blood serum with the suspect antigens. The antibodies crosslink the antigens and a precipitated sediment forms at the bottom of the tube. A precipitate in the test-tube indicates the presence of antibody that can bind that antigen (– no precipitate; ± trace precipitate; +++ strong precipitate). In the example antigens **D** and **G** have shapes (termed epitopes or determinants) which are similar to **A**, they are said to cross-react with **A**. (Also see Table 3.1 and Figure 3.6.)

of time, before their targets came into existence, seems hardly credible.[4]

Indeed it is *not* credible in terms of conventional Darwinian natural selection and survival of the fittest individual genes encoding a particular specific antibody.

71

Thus it was evident by the 1930s that the potential repertoire of specific antibodies had to be enormous and of the order of a million specificities or more. But the answer to a fundamental question was still not clear: How could a system evolve to make antibodies that were never a part of the evolutionary history of the animal? If infectious diseases have been instrumental in the evolutionary development of the modern immune system, then clearly the detailed and specific molecular characters of the many possible antigens have not been the driving force. Rather, it appears that a biological strategy to make an immune response to the unexpected has arisen over evolutionary time.

EVOLUTIONARY DEVELOPMENT OF THE IMMUNE SYSTEM

Given this dilemma, we start with the self-evident truism that infectious diseases have been the main, if not the only, selective force in the evolution of the vertebrate immune system. Indeed, we can quite easily envisage this evolutionary process in terms of the Darwinian 'survival of the fittest' immune system! If this is true, then the immunoglobulin (Ig) genes which are responsible for encoding antibodies are transmitted to progeny via the germ cells and their preservation throughout evolution is dictated by the same Darwinian rules of fitness which apply to other genes. Here we will critically consider this supposition.

Detailed comparative studies amongst both invertebrate (insects) and vertebrate species show that the adaptive immune system we have been discussing is present in cartilaginous fish (sharks and rays) and therefore arose at least 400 to 500 million years ago. These fish possess genes related to Ig variable (IgV), or T cell receptor (TCR) genes.[5] Studies by Robert Raison (University of Technology, Sydney) and others have established that vertebrates more primitive than cartilaginous fish such as the jawless fishes (hagfish and lamprey) do not possess an adaptive immune system; they do not possess IgV or TCR genes. The search is still on for the missing link. But at this stage, there are no known examples of the evolutionary steps between the jawless and cartilaginous fishes. In fact, there is no guarantee that the missing links will ever be found as they may now be all extinct. Figure 3.4 shows the main vertebrates in which the cellular and molecular structure of the immune system has been studied. Even in the cold-blooded vertebrates—the fishes—the main elements of an adaptive immune system of higher warm-blooded vertebrates can be found. All these immune systems have the capacity to:

- make an enormous repertoire of antibodies and T cells, thus facilitating a response to virtually any antigen;
- mount an enhanced (memory) response when challenged by an antigen a second time; and
- maintain self-tolerance.

When we are immunised with whooping cough or tetanus vaccines, our immune system 'remembers' the encounter with the antigen, allowing us to respond faster and more vigorously by producing higher concentrations of antibodies in the blood if the same antigen is encountered again. These features of antigen-driven somatic adaptability, specificity, self-tolerance and memory are the hallmarks of all vertebrate immune systems. In sharks and other cold-blooded animals, the low environmental temperatures cause much slower immune responses compared to the faster responses in the warm-blooded terrestrial vertebrates.

73

Figure 3.4 The vertebrates in which studies have established the presence of mammalian-like immune systems

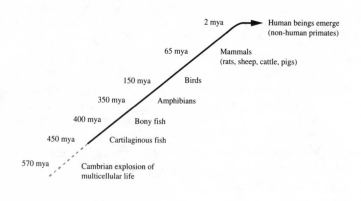

2 mya — Human beings emerge (non-human primates)

65 mya — Mammals (rats, sheep, cattle, pigs)

150 mya — Birds

350 mya — Amphibians

400 mya — Bony fish

450 mya — Cartilaginous fish

570 mya — Cambrian explosion of multicellular life

mya = million years ago

Most of our current information about the cellular and molecular behaviour of the immune system has come from experimental immunology involving immunisation of the inbred laboratory mouse. We also have much data from clinical observations on humans and we now know that virtually every cell, molecule and gene discovered in the mouse has a counterpart in the human. Although far fewer studies have been done on the lower vertebrates such as the cartilaginous fish and higher vertebrates (such as chickens, rabbits, and other domestic animals), it has become apparent that in the transition from cold-blooded cartilaginous fish to the warm-blooded land animals, the immune system has grown more complex. For example, we know that the genetic strategy used to store the long DNA sequences of information encoding the large number of different antibodies is uneconomic in the shark compared with that seen in mice and humans. Rabbits make even more efficient use of the DNA sequences. But chickens and other birds have evolved the most efficient system defined to the present time, as will become clear as we discuss the unique nature of the genes that encode antibodies.

STRUCTURE OF ANTIBODIES

Let us now briefly look at the structure of an individual antibody molecule. All antibody molecules are multi-sub-unit proteins with a common structural plan. Figure 3.2 shows the structure of the most common antibody called IgG. This basic structure was first described during the 1960s by groups led by Gerald Edelman and Rodney Porter, who shared a Nobel Prize in 1972. The antigen-binding site is composed of the complementary folding of the variable regions (V) provided by heavy (H) and light (L) protein chains called an HL heterodimer. There are two identical HL heterodimers per antibody molecule, except for pentameric IgM which has ten HL heterodimers (see Figures 3.2 and 3.5). The constant (C) regions of the molecule trigger the lysis (destruction) or phagocytosis (uptake by phagocytes) of foreign bac-terial cells and particles once the antibody has bound its target antigens (see Figure 3.1).

In terms of conventional genetics, one gene would be needed to encode an antibody H chain, and a second gene would be needed for the L chain. Is there sufficient length of DNA in our genome to encode all of the different antibody specificities (say a million)? This was an important question posed in the 1960s by Melvin Cohn and others, as the Genetic Code was being deciphered. Indeed this question, and the whole idea of the mechanism of self-tolerance, forced attention on to what type of strategy the immune system must employ to generate the diverse repertoire of antigen-binding

Figure 3.5 Diagram showing the protein structure of a pentameric IgM antibody molecule consisting of 10 HL heterodimers (5 monomeric IgM molecules)

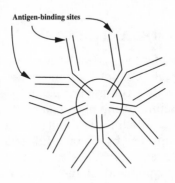

Antigen-binding sites

For further detail see Table 3.1. Each antigen-binding site may be of low affinity, but multipoint binding increases the overall binding avidity of the antibody (see Figure 3.7).

receptors required to fight off infectious diseases. Is it solely a 'germline' strategy whereby all antibody specificities are encoded in the egg and sperm cells? Or does the immune system employ a special somatic strategy in lymphocytes, in which the genes are mutated or randomly recombined to generate additional diversity in the repertoire of HL antibody-combining sites?

An often volatile conceptual battle between these two fundamental beliefs raged among scientists throughout the 1960s and early 1970s. However, by the late 1970s the problem was largely solved through the molecular genetic work of immunologist Susumu Tonegawa, who was awarded the Nobel Prize in 1987. We now know that some diversity is present in germline genes, but random somatic processes (recombination and mutation) are indeed essential for the generation of the enormous diversity of antibodies and T cell receptors. That is, during our lifetime our bodies 'learn' to fight numerous invaders and generate many new antibody-coding sequences within lymphocytes. Later we will discuss the evidence that indicates that gene sequences from lymphocytes may be incorporated in germ cells and passed on to the next generation.

We will now flesh out the essential elements of the antibody repertoire problem and the molecular details of the answer. Each antibody heavy chain is a protein constructed of about 400 amino acids (100 for the V-region, and 300 for the C-region), and each light chain consists of about 200 amino acids (100 for the V-region, and 100 for the C-region). Since each amino acid is specified on translation by a codon of three bases (see Appendix) this amounts to at least 1800 (600 × 3) bases of DNA sequence information necessary to encode each antibody HL heterodimer, if each H and L chain is encoded by a conventional gene. If there are 1 million possible different antibodies—a reasonable estimate based upon our current knowledge—then this means that the human genome would need to devote almost 2 billion (1800 × 1,000,000 or 1.8×10^9) bases of DNA sequence information just dedicated to encoding the possible repertoire of antibodies. Even if we just restrict the calculation to the essential encoded information, that is, to the critical V-regions which actually comprise the antigen-binding site of the molecule, then this number reduces to 0.6×10^9 bases. Yet the whole human genome has a maximum of only about 3–4 billion bases! At a bare minimum, then, somewhere between one-sixth and one-half of the DNA sequence space would have to be devoted to encoding antibody molecules!

77

When these numbers first became apparent about 30 years ago, the more discerning scientists realised that the Ig genes which encode antibodies must have an information storage strategy different from the other genes. At this point we will call all these other genes 'single-copy' or simply 'housekeeping' genes. Furthermore, it was becoming evident from other research conducted during the 1970s that there were a lot of apparently unused non-coding DNA sequences in the spaces between the genes on a chromosome, often referred to as 'junk' DNA. It was conservatively estimated that junk DNA comprised about 98 per cent or more of the total DNA in the vertebrate genome. ('Junk' DNA does not encode protein sequences, but the intervening stretches of DNA base sequence between and within the genes often play some functional role. This can be in the control of the structural folding of chromosomes, or in regulating the activity of genes and DNA replication.) These calculations meant that if conventional genes were used to encode antibodies, then there was sufficient coding DNA in the germline for only about 10,000 antibodies, rather than the 1 million estimated to be in the repertoire. And what about all the thousands of conventional housekeeping genes?

Thus, it was evident to many by as early as the late 1960s that a pure germline theory for the encoding of antibody genes was a patently absurd supposition. There had to be far more to one's life than just making antibodies! One possible answer was that diversity in the Ig genes responsible for encoding antibodies was somehow generated by somatic events. Susumu Tonegawa's demonstration of the unique rearrangement of DNA encoding V-regions of antibodies, and the work of many others showing somatic mutation in V genes, has finally solved the problem. This discussion is taken up in our section on the special structure of Ig genes on page 113.

WHAT HAPPENS WHEN WE BECOME INFECTED?

The immune system has evolved to combat acute infectious diseases. The outcome of infection depends upon a race between the multiplication of infectious agents and multiplication of the T (thymus-processed) and B (antibody-producing) lymphocytes of the immune system which provide the specific mechanisms for the elimination of the infectious agent. Let us now briefly consider the role of some of the actors involved in this battle and what happens when we become infected.

Antibodies belong to the class of soluble proteins called immunoglobulins (abbreviated as Ig), and are produced and secreted into the blood and tissue fluids by white cells called B lymphocytes. Antibody elimination mechanisms can involve auxiliary factors which allow direct lysis (destruction) of antibody-coated bacterial cells. Antibodies can also neutralise virus particles by binding strongly to them, preventing them entering or infecting target cells. Additionally, mobile scavenger white cells (called 'phagocytes') can engulf antibody-coated particles (phagocytosis) and destroy them. This process is depicted in Figure 3.1. It is advantageous if the antibody can bind at very low concentrations of antibody and antigen, that is, in the early stages of an infection when there are only a small number of disease-causing organisms present. This means that protective antibodies possess strong binding power or high avidity. If they are of low avidity, binding may not take place at low concentrations of either antigen or antibody.

In the case of disease-causing viruses which replicate within body cells, our immune system also responds by producing what are called 'killer cells' (cytotoxic T lymphocytes). Each T lymphocyte bears on its surface membrane a specific 'antibody-like' receptor which is

79

called the T cell receptor (TCR) which specifically recognises and binds to the molecular changes on the surface membrane of the virus-infected cell leading to the lysis of the infected cell. T killer cells do not harm normal uninfected cells because they do not recognise and bind to them via their TCRs.

Many people are not fortunate enough to enjoy the benefits of having an ideal immune system. In fact, there are many among us who learn to live with chronic immune problems. Some suffer from asthma or allergies as a result of the immune system overreacting to some environmental irritant. Others live with multiple sclerosis due to the immune system destroying their own nerve fibres or suffer from immune deficiencies which make it more difficult to fight off infection.

By the late 1960s, emerging evidence from the study of natural human immunodeficiencies showed that there were two basic categories of acute immune defects. Children without the ability to make antibodies (called 'agammaglobulinaemics') were unable to eliminate common bacterial infections, but recovered normally from the acute viral infections of childhood such as measles. In contrast, children lacking a normal thymus (called 'athymic') made antibodies and satisfactorily dealt with bacterial infection, but were unable to recover from viral infections. In the late 1960s Bob Blanden, working at the John Curtin School of Medical Research in Canberra, showed experimentally that recovery from acute viral infection in mice depended upon a T lymphocyte response against infection. By the early 1970s Bob, together with others in his laboratory, had discovered that T lymphocytes can kill virus-infected cells before the viral replication cycle is completed, thus nipping cellular infection in the bud. Other workers showed that T lymphocytes could produce soluble factors (called 'interferons') which have the effect of preventing viral multiplication in surrounding cells. Interferons, and other soluble cytokines produced by activated T lymphocytes, are now manufactured via gene cloning and expression of the cytokine genes in bacterial hosts. The pure cytokines are given to patients suffering from certain immune deficiencies to boost bone marrow and T cell activity and hence improve the response to infection.

In 1973, Peter Doherty and Rolf Zinkernagel, also working at the John Curtin School of Medical Research, showed that T lymphocytes recognised virus-infected cells in a way which involved not only the viral antigens (which was expected), but also the major transplantation or histocompatibility antigens of the cell itself (which was unexpected). Transplantation antigens are molecules responsible for rejection of foreign tissues. They differ from one individual to another and provoke strong T lymphocyte responses. However, grafting of foreign tissues is not natural, so the real biological function of the transplantation antigens was unknown, until the Doherty and Zinkernagel discovery. Cellular immunology was revolutionised and Doherty and Zinkernagel won the Nobel Prize in Physiology or Medicine in 1996.

During infection, antibodies are not normally made which can bind the combination of viral antigen and cellular transplantation antigen recognised by antigen-specific receptors on T lymphocytes (TCRs). In fact, if experimental antibodies are made which do bind to the site recognised by killer T cells, those antibodies, when added to a mixture of virus-infected target cells and killer T cells, can block the killing and allow viral infection to proceed successfully. This simple experiment illustrates a selective advantage for vertebrates with an immune system in which T cells and antibodies do not compete for binding to the same site on an infected cell surface.

The natural human immunodeficiencies described above showed that antibodies can be made against bacterial infection in the absence of a thymus (and the

T lymphocytes which develop in the thymus). These antibodies are of an immunoglobulin class called IgM. The secreted form of IgM made in response to bacterial infection without help from T lymphocytes is pentameric. That is, it consists of a polymer of five molecules of the structure shown in Figure 3.2, each of which has two binding sites for antigen. Thus pentameric IgM has ten binding sites for antigen (Figure 3.5).

An important axiom of biology is that function depends upon binding or sticking together of different molecules. In the case of antibodies and antigens, an interaction between a single antibody binding site and its complementary antigen site can be defined quantitatively by mixing different concentrations of the antigen and the antibody in a test-tube and determining how much of the interacting mixture is in the combined form (as antigen–antibody complexes), and how much is in the single, uncombined form. This is shown in Figure 3.6. This approach gives an assessment of what is called the 'affinity' of the antibody-combining site for its antigen. When more than one antibody-combining site is linked together so that you have two sites linked together in IgG or monomeric IgM (see Figure 3.2) and ten such sites in pentameric IgM (see Figure 3.5), a new factor comes into play. If the antigen is in the form of a particle with multiple identical antigenic sites displayed on its surface (e.g. a bacterium or a viral particle), then as soon as one antigen–antibody binding interaction occurs, the probability of a second combination increases and so on (see Figure 3.7). Thus pentameric IgM can successfully stick to the bacterial surface even if the affinity of an individual antigen-binding site is low (because of a relatively poor fit with the antigen). This multipoint binding of antigen with antibody is called 'avidity'. Thus, the high avidity of an IgM confers a selective advantage on the vertebrate host when it is

Figure 3.6 Stick diagrams of antigen–antibody complexes

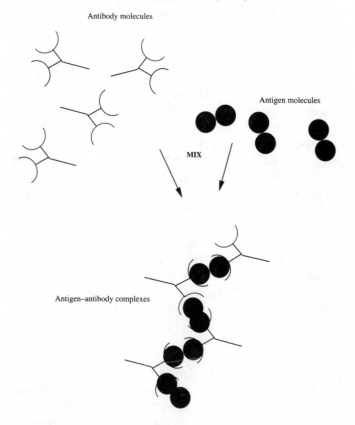

Antibody molecules

Antigen molecules

MIX

Antigen–antibody complexes

Complexes of antigens and antibodies are often insoluble and precipitate out of solution (Figure 3.3). Such complexes bind to the surface membrane of cells such as phagocytes via the constant portions of the antibody H chains enabling their engulfment and digestion (Figure 3.1). In Germinal Centres special cells called Follicular Dendritic Cells (FDCs) present such complexes for the selection of high affinity variable region mutants (see Figure 5.4). For other details see Table 3.1.

produced by B lymphocytes which proliferate in response to bacterial infection in the first hours and days, through its ability to bind to bacteria in the bloodstream and cause their elimination by various mechanisms.

Pentameric IgM is a large molecule that does not diffuse through the walls of blood vessels. A smaller antibody molecule is needed; this is IgG. The IgM response normally proceeds for a few days and then a switch occurs to the production of antibodies of the smaller IgG class which have only two binding sites instead of ten (Figure 3.8). This switch is achieved by a set of complex molecular mechanisms that involve specific DNA sequences in the Ig gene locus of the cells involved (and the molecular details are beyond the scope of this book).

The IgM-to-IgG switch illustrates an axiom in immunology, and one that is still not widely appreciated even by immunologists. This axiom states that the only evolutionary selective advantage for specificity in the adaptive immune system (antibodies and TCRs) is to enable discrimination between self and non-self. Thus, IgM is useful because despite low specificity and low affinity (poor fit) of an individual binding site, the ten binding sites on pentameric IgM enable the same antibody to bind to a variety of related antigens on different infectious agents. But the need for a smaller, diffusable antibody molecule has led to the evolution of a mechanism to switch antibody production to IgG. The two antigen-binding sites of IgG need higher affinity than the ten sites of pentameric IgM. The B cells selected for the switch process are those with the highest affinity for antigen, based upon competitive binding of antigen in specialised areas of lymphoid tissues (see Chapter 5). Another selective advantage of the switch mechanism depends upon the evolution of other classes of antibodies

Figure 3.7 Diagram showing the binding of pentameric IgM to a multiple array of the same antigen

Pentameric IgM – simplified to show 5 HL heterodimers. The actual molecule consists of 10 HL heterodimers (or 10 antigen-binding sites).

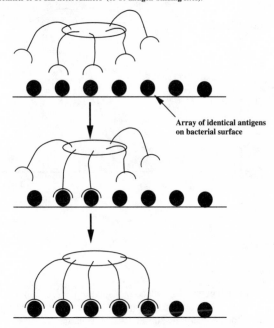

Array of identical antigens
on bacterial surface

Pentameric IgM has multiple identical antigen-combining sites each of which may have low binding affinity (see Figure 3.5). However pentameric IgM can bind very strongly (display high avidity) by binding in a multipoint fashion to antigenic surfaces such as bacterial cells (which present arrays of the same antigenic epitope). High avidity multipoint binding is depicted in the figure. As each antigen-binding site binds to its antigen, the probability increases for binding of a neighbouring site. When multiple antigen-binding sites are engaged, the total binding strength (or avidity) is much higher than a single binding site.

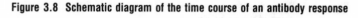

Figure 3.8 Schematic diagram of the time course of an antibody response

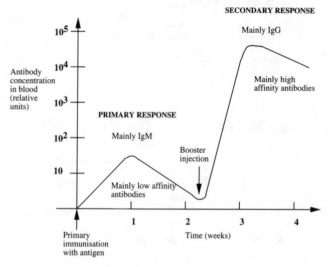

Plot of a typical time course of an antibody response in a laboratory mouse. In practice in humans 'booster' inoculations are usually separated by months or even years. The 'primary response' usually consists of lower affinity antibodies than those produced in later 'memory responses'. The antibodies made in the secondary and later memory responses are usually somatically mutated IgG molecules which are of higher binding affinity than those IgM or IgG antibodies produced earlier in the response (see Table 3.1 and Chapter 5 for more detail).

(IgA, IgE) with specialised properties, enabling them to function in particular physical locations in body tissues.

Our immune system is capable of recognising and responding to a large number of different antigens. The most common types of antigens are proteins and carbo-hydrates. The latter are polymers of variable numbers of glucose subunits and other sugar molecules. Consequently, the universe of potential biological antigens which can be formed is enormous. The immune system is characterised by its capacity to make an almost limitless number of specific antibodies against foreign proteins and carbohydrates. Each antibody is capable of distinguishing the foreign antigens from self antigens and contributing to the elimination of foreign antigens from the body.

In contrast with this virtually limitless repertoire of antibodies, T lymphocytes recognise only a short fragment (termed a 'peptide') of foreign protein which is bound in a special groove in a self-transplantation MHC antigen on the surface of a cell. T lymphocytes have a potentially large repertoire of receptors (TCRs), but the TCRs on functional lymphocytes are restricted to recognising variations in transplantation antigens on cell surfaces created by the binding of the short foreign peptide in the groove of the transplantation antigen. A complicated series of selection processess takes place during development of T lymphocytes in the thymus to ensure that this restricted repertoire of TCRs exists in the functional T lymphocyte population. Another complicated series of processes involved in the induction of antibody production in B lymphocytes ensures that individual animals do not produce antibodies against altered self-transplantation antigen with foreign peptides in the groove. Thus, antibodies and TCRs recognise non-overlapping aspects of the universe of foreign antigens.

How is this remarkable capacity to recognise a whole universe of different foreign antigens achieved while avoiding responses against self antigens? This question has now been answered following some 40 years of immunological research, and will be taken up in our discussion of 'clonal selection' in the next chapter.

87

THE NEED FOR SELF-TOLERANCE

We now diverge from the numerical and genetic problems posed by molecular diversity in the immune system to deal with the system's other fundamentally important feature, the tolerance of self.

As well as responding to the world of foreign antigens (the 'non-self' world), the other side of the coin in normal healthy individuals is that the immune system almost never makes antibodies or killer T cells against the organism's own tissues and cells. Normally, only some transient anti-self responses can be observed. This is despite the fact that the chemical composition of the proteins and carbohydrate polymers of the body are very similar to the molecular structures of foreign antigens. This failure to respond to self is called, appropriately enough, self-tolerance. It was defined at the turn of the twentieth century by the German microbiologist Paul Ehrlich who sketched its meaning in terms of the drastic consequences that would follow if self-tolerance were to break down. He called the phenomenon of the body's immune system not being able to differentiate between self and non-self 'horror autotoxicus'. Furthermore, we now know what Paul Ehrlich meant when our immune system runs out of control and self-tolerance breaks down. We develop debilitating autoimmune diseases when the body tries to reject itself, such as rheumatoid arthritis (caused by an immune response against joint tissues) and multiple sclerosis (caused by immune attack against nerve tissues). With respect to anti-nonself responses, we also know that the immune system is essential for survival because, when it is rendered ineffective (by congenital or acquired immune deficiency syndromes), humans and other vertebrates succumb to infectious agents that are normally controlled and eliminated by an immune response.

Yet, how is self-tolerance by our immune system achieved? Once set up, how is it maintained? Is it specified by our genes and thus preordained by the antigens and antibodies inherited from our parents? Or, is it acquired ('learnt') during the lifetime of each individual? These types of fundamental questions about the mechanisms of our immune system are at the heart of the perennial 'Nature' vs 'Nurture' debate. Scientifically and philosophically, they are not trivial. They are only now being resolved at the molecular and cellular level with respect to the immune system. They are also central to developing our understanding of how we might encourage our body's immune system to break normal rules and to fight cancer (self) or tolerate foreign organs and tissue grafts (non-self).

RESPONDING TO THE UNEXPECTED

The capacity to respond to the unexpected together with memory is shared by only one other somatically adaptive system, the central nervous system.[6] The key distinguishing features of the immune system are that the cells (the lymphocytes, phagocytes, and other white blood cells) can be both mobile and circulate throughout the body (Figure 3.9), or be part of a sessile (fixed) tissue system in the liver, spleen, lymph nodes, skin and intestinal tract. By contrast, in the central nervous system all neurons and nerve fibres are produced early in life and then retain their fixed, connected positions in space throughout adult life. However, like the nervous system, which monitors the status of body tissues as well as issuing instructions for muscle movement in all parts of the body, the immune system monitors the antigenic integrity of the body via a system of mobile cells which migrate through lymphatic ducts and blood vessels (Figure 3.9).

Figure 3.9 Human lymphoid system

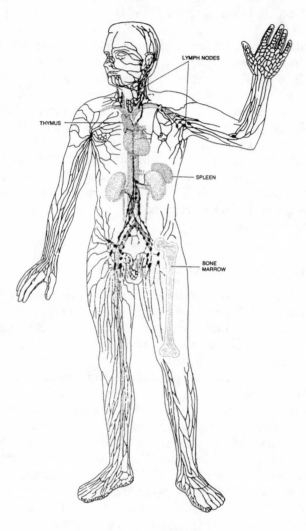

The system consists of an extensive network of lymphatic vessels and lymphoid organs (lymph nodes, spleen, thymus). The vessels are present in most tissues and organs, connect the lymph nodes, and finally drain via the thoracic duct into the blood. All B and T lymphocytes (and other white blood cells, including red blood cells) originate in the bone marrow (stem cells in the marrow are constantly dividing, producing millions of progeny blood cells each day). Those destined to be T cells migrate to the thymus gland via blood where they mature and are exported as mature T lymphocytes capable of T helper or T killer functions. The B lymphocytes which produce antibodies develop first in the bone marrow then migrate via blood and lymph ducts through the lymphatic system and populate the peripheral lymph nodes, such as adenoids, the tonsils and spleen. The lymph nodes which line the small intestine are called Peyer's patches. Mobile T and B cells can migrate to all parts of the body serviced by blood and lymphatic vessels, including the reproductive organs. Each lymph node is serviced by very fine blood capillaries which allow movement of molecules and cells from blood to lymph. (Adapted fom N.K. Jerne *Scientific American* vol 229: 52, 1973. Copyright © 1973 by Scientific American Inc. All rights reserved. We acknowledge the permission of the estate of the artist Bunji Tagawa to republish this figure.)

There is a great flux of cells and molecules (antibodies and other proteins) in both time and space.

In adults, all white blood cells originate in the bone marrow. Millions of new white blood cells are being produced on a daily basis. An enormous turnover of bone marrow-derived cells is required during the lifetime of an individual to continually replace those that die in various tissues around the body. The lymphocytes produced in the bone marrow then differentiate into antibody-producing B cells and thymus-processed T cells. The T cell precursors leave the bone marrow and enter the thymus, where they undergo further development (called 'differentiation'), including the expression of TCRs on their surface membrane, and exit as mature T cells.

Mature B and T cells (along with phagocytes) then populate the peripheral system of immunological organs such as the spleen and lymph nodes and move around the body via blood and lymph. The lymph nodes are strategically located all over the body and those draining the site of entry of an infectious agent (Figure 3.9) are usually the ones to react first. The immune system can therefore rapidly detect the entry of invading bacterial cells or virus particles. The nodes become enlarged (even tender) as the antigen-activated lymphocytes and other white cells divide rapidly there, or immigrate into the node from the circulation. We have all experienced these local and painful lymph node swellings (typically in the throat, armpit or groin) following vaccine inoculations or natural infections.

In the 1930s and 1940s a small band of immunological researchers and chemists (led by Frank Macfarlane Burnet, Linus Pauling, Karl Landsteiner, Friedrich Breini and Felix Haurowitz) were preoccupied with the question of how antibodies could be made so that the system could respond to the unexpected. This was long before we had knowledge of the chemistry and structure of genetic material (DNA), or the laws governing the flow of genetic information (i.e. DNA $\leftrightarrow$ RNA $\rightarrow$ Protein). During these years, the popular explanation involved the use of the mechanisms evoked by what was known as the 'instructionist' model of Breini, Haurowitz and Pauling. Put simply, it was believed that an antigen template could cause the specific folding of a naive, malleable, multi-potential Ig-protein into one particular shape, thus producing many copies of the specific antibody required to fight the infection. This is illustrated in Figure 3.10.

Such instructionist models are now known to be wrong. Proteins such as antibodies do not simply fold into a specific shape directed by the antigen acting as a template. We now know that the sequence of amino acids is what prescribes the folding pattern of the protein—with the assistance of recently discovered proteins called 'chaperones'. In turn, the amino acid sequence is determined by the linear nucleic acid base sequence of the DNA gene with its triplet code, via

Figure 3.10 Instructionalist model of antibody formation by Linus Pauling (1940)—the antigen template theory

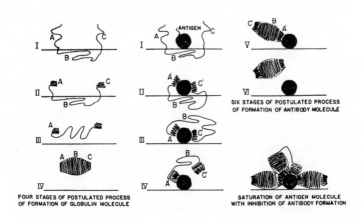

FOUR STAGES OF POSTULATED PROCESS
OF FORMATION OF GLOBULIN MOLECULE

SIX STAGES OF POSTULATED PROCESS
OF FORMATION OF ANTIBODY MOLECULE

SATURATION OF ANTIGEN MOLECULE
WITH INHIBITION OF ANTIBODY FORMATION

Diagrams representing four stages in the process of formation of a molecule of normal serum [immuno]globulin (left side of figure) and six stages in the process of formation of an antibody molecule as the result of interaction of the [immuno]globulin polypeptide [or protein] chain with an antigen molecule. There is also shown (lower right) an antigen molecule surrounded by attached antibody molecules or parts of molecules and thus inhibited from further antibody formation. (From Fig. 1 in Linus Pauling 'A Theory of the Structure and Process of Formation of Antibodies' *Journal American Chemical Society* vol 62: 2643–2657, 1940. Reprinted with permission of the American Chemical Society.)

messenger RNA. This is the Central Dogma explained in Chapter 2 and shown in Figure 2.4. However, features of the immune system, such as its capacity to exponentially multiply the concentration of antibodies during an immune response, its capacity to 'remember' previous responses, and its ability to ensure self-tolerance could not be explained by instructionist theories. Eventually, this led to the demise of 'instructionism' and permitted alternative 'selective' theories to emerge. The culmination of this is the Clonal Selection Theory of Burnet, first published in 1957.

CHAPTER 4

THE IDEA OF 'CLONAL SELECTION'

In 1900, Paul Ehrlich advanced what is now recognised as the first 'selective' theory of antibody formation. He predicted the existence of a 'multi-potential cell' (roughly like a B lymphocyte) which had the capacity to produce and express many different antibody molecules on its surface membrane. As a specific subset of cell surface antibodies bound to an invading antigen, this somehow stimulated the cell to produce additional, identical antibodies. This 'selective' process, occurring in many individual cells, would result in mass production of antibodies of that specificity. Ehrlich's imaginative idea is now known to be wrong. Modern 'selective' theories based on the idea that one cell can produce only one type of antibody (rather than multiple types) began to emerge in the 1950s.

The first significant modern selective theory was published by Nils Jerne in 1955. While he got it wrong by choosing the antibody molecule as the basic unit selected by an antigen, his contribution was seminal. He shifted the focus of immunologists away from the instructionist view (that antibodies were capable of folding into any shape, depending on the antigen template) and towards selective theories. In 1957, Macfarlane Burnet asserted that the cell is the basic unit selected by an antigen, and that one cell is responsible for producing one type of antibody. Burnet coined the term 'Clonal Selection'.[1] The central idea is that 'one cell makes one antibody'; many different cells (lymphocytes) exist and each one, when selected by the antigen which fits the shape of the antibody, multiplies to give a clone of identical cells all producing the same antibody. Clonal Selection Theory says that there is a mechanism which ensures that only one antibody is expressed per cell and excludes all others. We now know that the 'decision' to express one antibody per B cell takes place during early lymphocyte development.

David Talmage published the idea of cellular selection in a brief statement in 1957, but it is to Burnet that we owe the comprehensive development of the theory, including the marshalling of the pre-existing, supportive, experimental evidence and articulation of the implications of the theory for the self versus non-self discrimination problem. The Clonal Selection Theory was further developed by the efforts of researchers such as Melvin Cohn and Alistair Cunningham. It is still an important underlying view of how the immune system adapts to its highly variable and ever-changing antigenic environment. Figure 4.1 shows the details. Its main points have been verified experimentally.

The beauty of the theory was that it predicted how self-tolerance might be achieved. The rationale goes something like this. If a developing immature lymphocyte expressed an antibody receptor on its surface that bound to a self antigen, it received a 'negative' signal and was deleted (by death) from the immune repertoire. (In contrast, as stated above, mature lymphocytes would multiply, produce and secrete antibodies when antigen-bound to their antibody receptors.) Since self antigens are the first molecular shapes that the developing, immature lymphocytes are exposed to, this purging process that enables self antigens to be tolerated would occur at

Figure 4.1 Clonal Selection

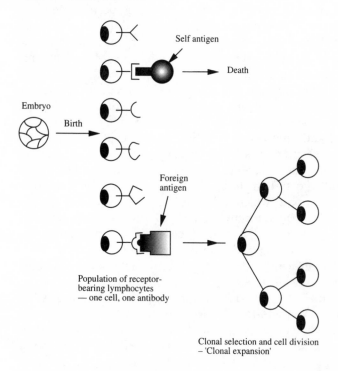

Other versions of this idea are outlined in Figure 1.2 and Table 3.1

the sites of lymphocyte development. Burnet referred to this as the deletion of 'forbidden clones'. It provided an elegant theoretical explanation of the self-tolerance problem. The lymphocytes surviving this selective filter (deletion of 'forbidden clones'), were then the only ones left to reach maturity and become available to bind with

97

foreign antigens. Since the enunciation of the Clonal Selection Theory in 1957, there has been a flood of immunological research defining the different classes of lymphocyte (B, T cells) and their roles in immune responses. However, the core concepts of the original theory are still valid today.

Thus, at the heart of Clonal Selection lie simple Darwinian selection principles. The diverse array of receptor-bearing cells is assumed to pre-exist before their first antigenic encounter. Self and foreign antigens then play selective roles (respectively negative and positive) in shaping the repertoire of lymphocytes in the blood and lymphatic system. It assumes that there is only one focal event, upon which all selection depends: this is the binding reaction between the cell surface antibody of B cells (or the TCRs of T cells) with antigens of the appropriate shape. Those cell populations resulting from 'clonal growth' produce the specific antibodies and comprise the killer cells (T cells) required to fight off infection. Clonal growth can amplify cell numbers by many thousand-fold. This explains why the curve of antibody concentration in blood over time rises in an exponential fashion over the days and weeks following infection or artificial inoculation of foreign antigen (see Figure 3.8). The increase in antibody concentration reflects the exponential increase in the clonally selected population of antibody-producing cells which double their numbers by cell division every few hours. After fifteen divisions there would be over 30,000 cells derived by division of one original cell through the 2^n power series where 'n' equals the number of divisions. These large increases in numbers of cells secreting antibody enable a patient to produce sufficient antibody for it to reach an effective concentration to counter acute infections. (As a rule of thumb this takes about three to five days.)

It is easy to imagine how the Clonal Selection Theory can account for immunological memory. The explanation is based (at least in part) on antigen-driven multiplication of antigen-specific cells (clonal expansion). Once antigen-specific cells have multiplied, some of the progeny cells can be long-lived. These 'memory' cells can lurk in the circulation and lymphoid tissues—awaiting the next encounter with the same antigen many years later.

The theory is also useful in that it can provide a plausible explanation for the phenomenon of 'affinity maturation'. Late in an immune response, the antibodies made are of higher binding affinity than those made earlier (see Figure 3.8). Over time, as the antigen concentration in lymphoid tissues falls, competition between different B cells for binding to scarcer antigen results in the more successful B cells being 'selected' to reproduce. The B cells with surface antibody receptors of the highest affinity (antigens binding more firmly) will win the competition, and their high affinity antibody will eventually become dominant. More details of this intense affinity-based selection will be recounted in the next chapter (see Figure 5.4).

If we were able to examine each of the antibody proteins in a sample of our blood, we would find thousands of different antibody molecules—each capable of recognising a specific antigen in the environment. If we were to determine the amino acid sequence of each of these antibodies we would also show that each antibody differs from the next in the amino acid sequence of the variable (V) regions which form the antigen-binding site (see Figure 3.2). When this fact was established in the 1960s it provided a critical boost to the validity of Burnet's Clonal Selection Theory.

One important experimental test of Clonal Selection Theory was performed in the late 1960s by our colleague Gordon Ada working with Pauline Byrt at the Walter and Eliza Hall Institute of Medical Research, Burnet's scientific 'home'. The experiment was based upon the prediction that if a population of lymphocytes were exposed to a single antigen, those lymphocyte clones with antibodies reactive to the antigen would bind it while the vast majority would not. (They would have antibodies reactive to a variety of other antigens.) By making the antigen radioactive, Ada and Byrt reasoned that the lymphocytes which bound to it would be irradiated and become incapable of making an immune response. Thus, the lymphocyte population would be defective with respect to an immune response to the radioactive test antigen but perfectly capable with respect to any other antigen. This was indeed the result they obtained. The experiment became famous as 'the antigen suicide experiment' and was the first accepted experimental proof of the Clonal Selection concept.

The scientific acceptance and validation of the Clonal Selection Theory has served to focus Ted Steele's and Bob Blanden's research efforts on discovering the mechanisms underlying the evolution of antibody genes. This new era in immunology is uncovering the details of the complex genetics giving rise to the diversity and affinity maturation of antibodies. Since the late 1970s the extraordinary and unique molecular genetics of the Ig and TCR gene systems has been progressively revealed. We now know many of the details defining the special DNA sequence structure of these genes. We can now explain how one cell makes one antibody, and how new antibodies are generated by mutation, enabling the immune system to produce high affinity antibodies to the many thousands of possible unexpected antigens.

The Clonal Selection Theory focused attention on the question of whether Ig genes undergo mutation within individual B lymphocytes after stimulation by antigen. Burnet first posed the question in general terms in 1957.

But he was content to just consider that the various Ig genes encoding antibody specificity *pre-existed* before foreign antigens impacted on the system. (Darwin, in the previous century, also took for granted pre-existing genetic variation amongst animals and plants upon which 'natural selection' could act.) Burnet thus assumed that the antibody genes were clonally expressed in a diverse population of cells which could then be antigen-selected and clonally expanded. He also did not have access to the sophisticated technologies of modern molecular biology which are needed to answer the question of somatic mutation.

By the late 1960s and early 1970s, however, Melvin Cohn and then Alistair Cunningham began to formulate molecular explanations for why the immune system should be able to generate new somatic mutants of antibody genes in response to a foreign antigen. By taking this reasoning to its logical conclusion, it can be speculated that maybe only a small number of necessary Ig genes need to be propagated in the germline DNA, and that the rest could arise during the lifetime of the animal by somatic mutation stimulated by antigen. Yet, at the time these 'somatic mutation theories' were first being articulated, they were only appreciated by a small number of immunologists. The majority continued to hold the view that all antibody specificities were encoded in the germline genes of the species. That is, they pre-existed before the animal was born. This also presupposes that variability will be the result of chance (random) mutations in the germline genes which are passed on to offspring. This disunity among scientists led to one of the most fundamental and philosophically significant debates among immunologists. In fact, it could be argued that the *somatic* versus *germline* debate on antibody formation has been at the core of much of the most important work in molecular genetics this century, and

101

is on a par with the great battles in physics in the 1920s on the wave–particle duality in quantum mechanics. It is also the reason why we have felt compelled to explain the underlying mechanisms in the wider context of their historical roots. We will take up this theme again shortly.

WHY ANTIBODIES ARE SPECIFIC AND HOW SELF-TOLERANCE IS SOMATICALLY ACQUIRED

Let us now return to the problem of self being distinguished from non-self. We have already said that it is necessary for antibodies to be specific so as to ensure that the self versus non-self discrimination mechanism can work. Thus, to be able to distinguish between our own large universe of self molecular patterns, and those from the outside world (which have some of the same chemical characteristics), evolution has ensured that we produce a large, diverse repertoire of antigen-recognition molecules. Each specificity is clonally expressed in populations of thousands of specific cells and the 'clonal deletion' mechanism can simply edit out those cells which are self-reactive. This scenario emphasises the necessity for specificity and diversity of antibodies. By way of a simple but not strictly correct metaphor,[2] the picture definition on a TV screen depends on the number of pixels per scan-line across the screen. The more dense the pixel lines, the clearer the image. It is also necessary to have a diverse range of pixels (light, dark, red, green and so on) to ensure that it is easier to distinguish between the different sections and hence identify the resulting image. Thus by analogy, the less specific antibodies and T cell receptors are, the more they run the risk of being deleted by the self-tolerance process. If all of them were unspecific, then there would be two possible outcomes. In the first scenario, we would have

102

no immune system left after self-tolerance was imposed. Alternatively, if self-tolerance did not exist, our own immune systems would attack and destroy the bodies that house them.

Once the need for self-tolerance is accepted, it then becomes axiomatic that it is necessary for self-tolerance to be 'learnt' in a Darwinian manner in each individual. Let us explain. If males of inbred mouse strain **A** are mated with females of a genetically different inbred mouse strain **B**, then progeny are produced which inherit different genes from each parent. **A**-line mice are programmed by their genes to make **A** antigen molecules in cells and tissues, whereas **B**-line mice are genetically programmed to make **B** antigens. In a manner similar to the ABO blood transfusion rules in humans, **A** individuals as adults will always have developed anti-**B** immune reactivity (naturally developed antibodies and T cells) and **B** individuals will have developed anti-**A** reactivity. Thus a tissue transplant of **A** into **B** will always be rejected, and vice versa of **B** into **A**. But these facts raise a fundamental problem for offspring from genetically different parents which will be **AB** in their antigenic make-up. If self-tolerance were not 'learnt' during development, the **AB** offspring of the cross could not survive. They would have self-destructed as soon as their immune system was capable of mounting a response to their own bodies. We now know that normal human and most wild animal populations show great diversity in the genes encoding the transplantation antigens. Therefore all of us carry different combinations of transplantation antigens inherited from our parents. That is, we are all like the **A** × **B** cross illustrated in Figure 4.2.

Clearly then, the **AB** individual must have 'learnt' to tolerate both **A** and **B** antigens at some stage of its own development. The self-tolerance mechanism has to

Figure 4.2 Self-tolerance is not genetically determined

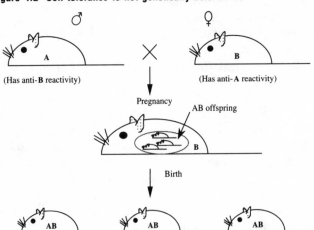

No anti-**A** or anti-**B** antibodies or reactive T cells in AB progeny
implying that self-tolerance is 'learnt' during development

Immunological theorist Melvin Cohn and his colleagues Peter Bretscher and Rod Langman have repeatedly emphasised this key point. The figure shows the outcome of an 'experiment of nature' (resulting from random breeding between genetically different individuals). Since mature **AB** progeny do not self-destruct in an autoimmune reaction, this clearly demonstrates that self-tolerance is 'learnt' by the individual during development. Macfarlane Burnet's Clonal Selection Theory explains how this 'learning', based on Darwinian negative selection within the immune system, is achieved—by the deletion of 'forbidden' anti-self clones (see Figure 4.1).

be 'somatically acquired'. During the early 1950s Sir Peter Medawar and colleagues added substance to this view of immune development by testing and proving a theoretical proposition first published by Macfarlane Burnet and Frank Fenner in 1949. Medawar's team showed that if **B**-line newborn mice were injected with

104

cells of the **A**-line, then they grew up to be tolerant of **A**-line skin grafts. Thus **B**-line mice could be somatically modified to enable them to permanently accept an **A**-line skin graft which they would normally have rejected as an adult. This is shown in Figure 4.3. Burnet and Medawar shared the Nobel Prize in 1960 for their important discovery of 'acquired immunological tolerance'.

Our discussion to this point has led us to conclude that the acquired somatic learning program necessary for both immunity and tolerance is based upon the key interaction (or binding) of antigens with clonally expressed cell surface antigen-recognition molecules (antibodies, TCRs). It is achieved in a Darwinian selection process involving diverse populations (clones) of lymphocytes. This is the central principle of the functioning of the immune system. However, what the discussions in the next two chapters will show is that it nows appears necessary to graft onto this view of immunology the concept of a 'Lamarckian gene feedback loop'. As we will explain, the soma-to-germline gene feedback process provides a coherent explanation of *all* the unique molecular genetic features of the immune system as we understand them today. But to take the reader further in this journey of explanation, we must now take time to explain some of the details of the DNA sequence structure of genes, and to highlight some of the unusual features of the Ig and TCR genes.

WHY ANTIBODY GENES ARE SPECIAL: VARIABLE REGION DNA REARRANGEMENTS

The paradigm of 'DNA makes RNA makes Protein', previously described in Chapter 2, applies for bacteria through to the more complex vertebrates such as humans. However, there is one important difference

Figure 4.3 Peter Medawar's demonstration of acquired neonatal tolerance

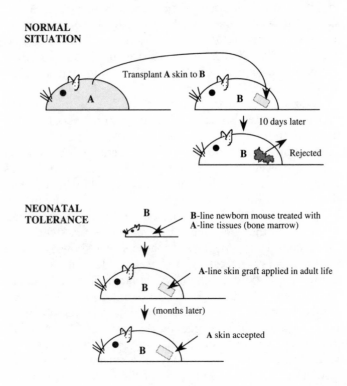

**NORMAL
SITUATION**

Transplant A skin to B

A

B

10 days later

B   Rejected

**NEONATAL
TOLERANCE**

B

B-line newborn mouse treated with
A-line tissues (bone marrow)

B   A-line skin graft applied in adult life

(months later)

A skin accepted

B

Peter Medawar and colleagues Leslie Brent and Rupert Billingham conducted
these seminal experiments in the 1940s and early 1950s. The experiments are
consistent with the predictions of Macfarlane Burnet's Clonal Selection Theory
(Figure 4.1) and demonstrate that introduction of foreign cells from bone
marrow during early neonatal development can induce a state of 'acquired
immunological tolerance': the early exposure to foreign antigen causes deletion
of the anti-foreign clones and the system is now 'tricked' into treating the
foreign tissue as 'self'.

106

between the structure of genes of higher cells, which includes vertebrate cells, and the corresponding genes in bacteria. The coding sequences of vertebrate genes almost always have inserts ('introns') of non-coding DNA sequences. The coding sequences are called 'exons'.

In the late 1970s, researchers R. Roberts and P. Sharp (Nobel Prize 1994) showed that the initial messenger RNA (mRNA) copy made in the cell's nucleus contained these intron sequences. But more puzzling was their discovery that by the time the mRNA was exported from the nucleus and passed into the body of the cell ('cytoplasm') the intron region was missing. It had somehow been deleted and the exons (protein coding sequences) were now joined perfectly so that the contiguous mRNA could be 'read off' in codon triplets into the specified protein.

The process of removal of the introns (non-coding sequences) is called 'RNA splicing'. Splicing is very precise; it rarely cuts the RNA message in the wrong place. We now know that there are specific RNA signal sequences recognised by a special enzyme complex ('spliceosomes') to clearly define the boundaries of the intron. Some introns are ribozymes (RNA enzymes) capable of self-splicing, possibly relics from the 'RNA World' of many billions of years ago. Many spliceosomes are complex enzymes of RNA and protein. The important point to make here, and it will be stressed again later, is that the splice sites ('cut-sites'), which are coded in the DNA sequence, are only cut by the spliceosomes operating on a single-stranded RNA sequence after transcription from the DNA. The double-stranded DNA in the genome is never cut at these sites. 'Single copy' genes as illustrated in Figure 4.4 are all those genes which encode proteins that are essential for 'housekeeping' functions of the cell or multicellular organism. These

Figure 4.4 A schematic outline of a bacterial gene and a 'single copy' gene in a higher cell

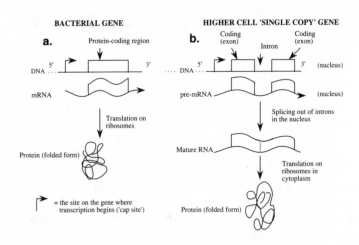

The diagram shows the different structures of a protein-coding gene in a bacterial cell in contrast to a protein-coding gene in a higher cell. Note that it has become accepted practice to refer to the left-hand, or upstream region of the DNA strand, as the 5′ end, and the right-hand side or downstream region as the 3′ end (pronounced 'five prime' and 'three prime' respectively). The term 'single copy' refers to the fact that there is only one functional copy of the gene per chromosome (to contrast with multigene V families, Figure 4.5, where there are *many* similar gene copies per chromosome, each with a high degree of DNA sequence similarity). The region of the DNA which codes for protein is shown by a rectangular box. Double-stranded DNA is shown by the straight lines, and single-stranded messenger mRNA is shown by the wavy lines. The hooked arrow indicates the start site for transcription of the mRNA.
a. In bacteria the protein-coding region is a continuous set of triplet codons each specifying an amino acid (see Appendix). The mRNA is transcribed and is immediately translated into the amino acid sequence (protein) on ribosomes.
b. In higher cells *almost all* protein coding genes are more complex, displaying amino acid coding stretches (called 'exons') interrupted by non-coding regions called 'introns' (indicated by the straight line gap between coding regions). The only 'intronless' exceptions in higher cells are the genes which encode

histones (Table 5.1) and the processed pseudogenes and functional retrogenes arising by reverse transcription of an mRNA molecule (see Chapter 7). Thus a long pre-mRNA molecule containing introns is copied from the DNA template strand. A special molecular organelle in the nucleus (termed a 'spliceosome') then identifies the exon–intron boundaries, cuts out the introns from the pre-mRNA and precisely joins the exons together to give a contiguous coding region as shown. This mature or 'processed' mRNA is then exported from the nucleus to the cytoplasm where it is translated into a sequence of amino acids on the ribosome (see Appendix).

genes are subject to life-or-death Darwinian selection decisions. For example, the genes coding for the protein subunits of the haemoglobin molecule which transports oxygen in red cells from the lungs to the respiring organs around the body are single copy genes. Faulty molecules caused by mutations are usually inefficient oxygen trans-porters and are consequently lethal or debilitating.

Genes in higher cells therefore have a type of RNA-editing mechanism that occurs before they are expressed in the form of a mature mRNA molecule which has been created to code for a specific and contiguous sequence of amino acids which makes up a protein. The editing out of the RNA introns and the joining together of the amino acid-coding sequences in the mRNA has to be precise to ensure that the reading frame of the base triplets is maintained (see Appendix). If an error is made here (and it rarely is) it may produce an mRNA with an 'out-of-frame' sequence. This most commonly leads to the appearance of an aberrant protein and premature termination of protein synthesis by what is appropriately called a 'stop' codon now appearing upstream of the normal position. (Stop codons do *not* specify an amino acid; thus they bring to an end the sequential addition of amino acids to the protein chain. The three alternative stop codons in an mRNA sequence are **UAG**, **UAA** and **UGA** (see Appendix).) These stop codons normally define the 3′ boundary or right-hand end of the coding sequence of a gene. Premature termination results in a shorter protein, usually with sub-optimal function. All normal protein-coding genes are therefore a contiguous stretch of triplet codons with a stop codon at the end (in the same way that a full stop defines the end of a sentence) and are termed 'open reading frames' (abbreviated as ORF). Darwinian evolutionary selection pressures would normally act against any mutant giving rise to a stop codon within the ORF.

The emerging story of the molecular genetics of Ig genes began in the mid-1960s. It was clear at this point that both heavy (H) and light (L) antibody chains had a basic functional dichotomy (see Figure 3.2). They consisted of an amino acid sequence called the variable or V-domain (for antigen binding) and a constant or C-domain (e.g. for triggering target cell lysis or phagocytosis). However, unlike all other genes located and mapped by geneticists to specified positions (termed loci) on the chromosomes, antibody genes presented a profound paradox. The genes encoding V- and C-domains seemed to be physically separated by vast stretches of DNA sequence. For example the V- and C-domains of heavy chains in the mouse are on chromosome number 12, down towards the end of the long arm near the end of the chromosome (the 'telomere'). Yet genetic mapping, and later DNA sequence studies, clearly showed that the V portion was at least 300,000 bases (written as 300 kb, or kilobases) upstream of the C portion! Why is this?

Even in the mid-1970s when the intron structure of conventional genes was being defined, this large separation was considered unusual. However, some ten years earlier, researchers Dreyer and Bennett predicted that the contiguous piece of DNA encoding an H or L protein chain is produced when the V portion of the DNA 'translocates'—or rearranges its relative position to come close to the C portion. That is, a specific type of DNA sequence editing or splicing had been predicted to account for this paradox. Further, they had also postulated that this process took place during the maturation of the lymphocyte. Dreyer and Bennett suggested that the upstream portion of the Ig locus (or site of location on the chromosome of the DNA sequences encoding Ig genes) encoded a large number (100s?) of 'germline' V genes and in any one single B cell, one of these V genes 'moves' and becomes integrated next to one of several possible C-region genes. By drawing only one V gene at random from a large number of different possible V genes, the cell could become committed to produce only one specificity of antibody. After this random rearrangement of one H and one L chain gene in a cell, the cell could then assemble a complete antibody protein of previously 'unknown' specificity from the large number of possible combinations of the H + L protein chains. The cell would then be able to express this antibody on its surface and the fate of the cell would be sealed: if it happened to be anti-self, it would be deleted. If it survived, it could become a potential player in immune responses to foreign antigens.

This theoretically appealing yet quite bold prediction of DNA rearrangement was proved in principle some ten years later by the work of Susumu Tonegawa, who revealed the organisation and rearrangement of genes in the Ig loci of mice.[3] The DNA rearrangement turned out to be even more complicated than predicted. This is explained in Figure 4.5, which applies to mouse and human. Other vertebrates can have different, yet similar in principle, organisation of their Ig-encoding genetic elements.

While there is still much research needed to track down, accurately map and sequence every V gene in the mouse and human genome, we now have a clear picture of what are known as the 'germline' and 'somatic' configurations of Ig genes. The scheme in Figure 4.5 is for Ig heavy chains, but very similar flow charts can be drawn up for Ig light chain genes and for TCR genes expressed in T cells. Notice that there are a hundred or so V genes (or V-elements) upstream of the smaller number of what are called 'joining' (J) and 'diversity' (D) elements and that these are separated from the small number (about eight) constant (C) region genes. The D- and J-elements are each responsible for encoding about three to fifteen amino acids whereas the V-element is responsible for encoding about 100 amino acids.

The separate genetic elements of antibody genes in the non-rearranged format are said to be in the 'germline configuration', i.e. in the same form as the DNA in the germ cells (sperms and eggs). This is also the form of the DNA in all those cells of the body that are not mature lymphocytes (e.g. cells of liver, kidney, pancreas and other white blood cells such as phagocytes). However, in maturing lymphocytes of mice and humans, the DNA undergoes random somatic rearrangement such that, in any one cell, one particular V-element translocates to join up with one D- and one J-element in a position just upstream of the first C gene. The C coding region itself is broken up by a number of introns. This still leaves a large intervening sequence between the somatically rearranged V gene, which we now give the generic name V(D)J gene, and the C-region. This is called the 'somatic configuration'. The bracket around the 'D' in the generic symbol 'V(D)J' arises because Ig light chain genes are built only from V- and J-elements.

In the germline configuration, the DNA in the chromosomes cannot be transcribed into mRNA that encodes a complete H or L protein chain, so that antibodies cannot be synthesised. In the somatically rearranged configuration, the rearranged V(D)J and C gene are actively

112

Figure 4.5 Schematic outline of the germline configuration and somatic rearrangement of V and C coding regions of antibody genes (heavy chain) in humans and mice

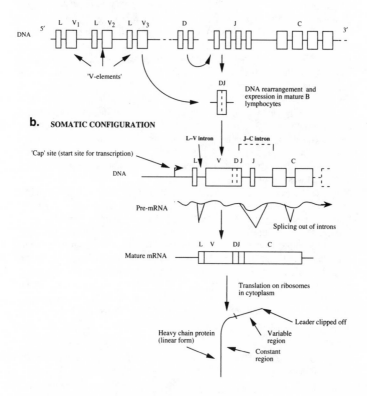

a. In the germline configuration, an array of very similar (yet distinctly different) germline variable gene (V) elements lie upstream of the constant (C) region coding exons. Such germline V-elements (encoding about 95 amino acids) are said to be *unrearranged*. In between the Vs and Cs are the diversity (D) and joining (J) region elements (each codes for a short stretch of amino acids). Each V-element has a leader (L) sequence which encodes a short stretch of amino acids called the 'signal peptide'. All genes encoding proteins such as antibodies destined for export (secretion) from the cell or to another membrane-bounded compartment within the cell have 'signal peptide' tags which allow them to be shunted to their correct cellular or extracellular locations. The intron between L and V is called the 'L–V intron' (also referred to as the 'leader intron') and its significance for understanding the DNA sequence structure of the 'soma-to-germline integration footprint' is outlined in greater detail in Chapter 6.

b. During B cell development in the bone marrow, each B lymphocyte randomly rearranges its heavy chain V, D, J-elements so as to give a rearranged variable region sequence, VDJ, now said to be in the 'somatic configuration'. For H chains, the first step in rearrangement is the random formation of a fused DJ, followed by the random fusion of an upstream V-element to create a VDJ. After rearrangement it is usual to find that all the intervening DNA between the V-element chosen by chance for rearrangement and the utilised J-element has been looped-out and deleted from the cell. Each successful variable region DNA rearrangement therefore is *unique* to each maturing B cell. In the figure, notice that the rearranged variable region (VDJ) lies upstream of the C protein coding exons. Note also that between the VDJ and C, termed the J–C intron, there is an unutilised J-element (which, in this example, has simply escaped being used by chance). The pre-mRNA containing the antibody variable region (VDJ) attached to the C-region is made in the nucleus. The L–V and J–C introns (together with the introns within the C-region) are then spliced out, resulting in the export of a mature, processed, mRNA molecule to the cytoplasm where it is translated into an amino acid sequence on ribosomes. The leader (signal peptide) is clipped off the protein as it is secreted from the cell. (Also see Tables 3.1, 5.1, Figure 4.4 and the text for additional explanations of terms.) The germline and somatic configurations for light (L) chain genes are very similar except that L chains do not have D-regions.

transcribed, resulting in the production of mRNA, which is translated into H and L chains to form an antibody. The somatic mutation machinery is also only active on V(D)J rearrangements and *never* on unrearranged V-elements in the germline configuration.

The DNA rearrangement process is executed by a set of special DNA recombination enzymes encoded elsewhere in the genome by what are called 'recombination activating genes' (RAG). The RAG proteins recognise DNA sequences (called recombination signal sequences) on the appropriate sides of V-, D- and J-elements which are joined together. The first step in heavy chain expression in a maturing B cell in the bone marrow of humans or mice is a 'DJ' rearrangement. All the intervening DNA between that D-element (there can be twenty or more of these) and that J-element (there are four or five of these) is removed. The next step is the random translocation of the DNA at any one of the hundred or so V-elements to the rearranged DJ, and again all the DNA between the translocating V-element sequence and the DJ is removed.

The Dreyer and Bennett proposal (see above page 111) theoretically reduced the amount of genetic material required to encode antibodies by postulating the multiplication and diversification of only those genes encoding the variable (or V) portions of immunoglobulin protein chains, while at the same time retaining one constant (or C-region) gene for each of the heavy (H) and light (L) chains. However, this strategy only effectively reduced the required coding DNA by about three-fold. Tonegawa's demonstration of the V(D)J rearrangement, coupled with the notion of random use of about 100 V genes, 20 D genes and 4 J genes, together with random use of heavy and light chains in heterodimeric antibody binding sites, provided for a system which was able to generate great potential diversity from a relatively small amount of genetic material. In addition, some of the joints between V, D, and J contain untemplated base additions and deletions, thus further increasing diversity.

With an understanding of these fundamental molecular mechanisms, the reader should now be in a position to be able to address two of the most important contemporary issues in immunology. What is the mechanism that ensures that a B lymphocyte makes only one antibody specificity? What is the mechanism which confines somatic mutation (in B lymphocytes) to rearranged V(D)J regions?

115

All somatic or body cells have a double set of chromosomes, one set from the mother, the other set from the father. This is said to be the 'diploid' state. In contrast, sperm and eggs have only one set of chromosomes (the 'haploid' state). Fertilisation, the fusion of sperm and egg, restores the diploid state. Apart from the sex chromosomes (X and Y), the other chromosomes in each diploid set match up into pairs with very similar DNA sequence. As a B lymphocyte matures, rearrangement of V-, D-, or J-elements begins. A successful rearrangement of a V(D)J in one chromosome sends a feedback signal that turns off RAG enzyme activity and prevents successful rearrangement in the second chromosome. This is the answer, in principle, to the first question on how a B lymphocyte makes only one antibody specificity. Further detail is beyond the scope of this book.

The second question on how mutations are confined to V(D)J regions is an essential part of the discussion of a soma-to-germline feedback loop which is the central theme of this book. It is discussed in detail in the following chapter. Here, we outline some important general principles. While mutations in variable portions of antibodies can be useful, mutations in the constant regions are potentially very dangerous because they could render an antibody useless in its effector functions, such as triggering the lysis of bacterial cells or the promotion of phagocytosis (see Figure 3.1). The separation of V- and C-region genes has allowed the evolution of mechanisms to mutate the V-region while preserving the C-region. This is what occurs in B lymphocytes at certain stages of their life. The rearranged V(D)J gene can be subjected to very high rates of somatic mutation and the resulting new antibodies are selected for utility in antigen binding (see next chapter). Those with the highest affinity for the foreign antigen win the antigen-binding

116

selection contest and the cells producing them are retained and become long-lived memory cells. Those mutations which reduce or eliminate affinity for the foreign antigen cause the B lymphocyte to lose the contest and such cells die. Those B lymphocytes encoding antibodies which bind to the self antigens are subjected to elimination by a special process, thus preserving self-tolerance.

We now have a very good understanding of how the immune system can encode millions of specific antibodies, and yet still have plenty of DNA sequence space left over for all the other genes required by the cell and multicellular organism. A simple set of calculations showing the germline potential can illustrate this point:

- In a mouse or human B cell, the germline configuration of H chains encodes about 100 V-elements, about 20 D-elements and 4 J-elements. Under a random rearrangement model this can generate $100 \times 20 \times 4 = 8,000$ possible H protein chains.
- The germline configuration of the main gene cluster specifying L chains encodes about 100 V-elements and 4 J-elements. Under a random rearrangement model this will generate $100 \times 4 = 400$ possible L protein chains.
- Thus the total number of possible HL protein heterodimers, assuming all H + L associations produce viable antibodies, is $8,000 \times 400 = 3.2 \times 10^6$, or about 3 million antibody specificities.

These calculations do not take account of additional diversity generated by untemplated base additions or deletions in the V(D)J joining process.

But how many of these 3 million potential specificities are ever used? The short answer is that we do not know. However, we do know that about one-half of the

117

V genes in germ cells have never been found to encode an antibody (that is, they have not been found to be functionally rearranged in a B cell). One theoretical factor which limits repertoire size functionally is the logic of the Protecton Theory of Rod Langman and Melvin Cohn.[4] A mouse has about 50 million B cells. If a random repertoire of about 3 million different antibody specificities was evenly distributed in about 50 million B cells, the average number of cells with a given specificity would be about seventeen. During acute bacterial infections, in order to win the race against rapidly multiplying bacteria the immune system must produce a sufficient concentration of antibody with sufficient binding avidity for the bacterial cells to cause elimination of more bacteria than those produced by multiplication. Since bacteria can divide in less than one hour and since B cells take at least five to six hours to divide, this race can only be won if a large initial number of B cells with avidity for a particular bacterium are present in the repertoire. Other critical factors are the rate of antibody protein production in B cells, and their physical location in relation to the site of infection. The closer they are, the higher the local concentration of antibody and the greater the rate of elimination of infection. These constraints have led Langman and Cohn to propose that the initial functional antibody repertoire (before somatic mutation) must be far smaller than the potential repertoire. They proposed a figure of 10,000 specificities, giving an average of 5,000 B cells per specificity in a mouse.

But all these estimates of the size of the functional antibody repertoire are still very large numbers. The core strategy in mouse and human immune systems is based on random rearrangement of V, D, and J genes. Fully functional antibody proteins are encoded in the germline as bits of DNA awaiting somatic rearrangement and assembly into a functional V(D)J gene (as depicted in Figure 4.5). The random combination of the H and L protein chains then gives an HL antibody heterodimer. It is an extremely economical genetic storage strategy allowing the encryption of millions of pieces of potentially useful protein information.

Additional somatic variation of rearranged V(D)J sequences in B lymphocytes also contributes to diversity in the repertoire by selective expansion of particular antibody specificities. We now know that antigen-driven somatic hypermutation of rearranged V(D)J regions in mature B cells is a fact. This is important because it enables selection and expansion of mutant B cells with antibodies of high affinity for a particular foreign antigen. There is also the newly recognised genetic recombination phenomenon described by the group led by Martin Weigert and others called 'V gene replacement'.[5] This is known to occur for heavy chains; another upstream V-element can precisely replace the existing V portion of the DNA sequence in the VDJ. The same B cell line may undergo several successive V gene replacements during clonal growth.

The other obvious point to reiterate about the V(D)J rearrangement and joining process is that the complete gene encoding the variable part of an antibody is assembled from several bits of DNA sequence. The multiple members in each family of bits (V, D and J) are assembled at random, and there is *rarely* any fully assembled contiguous DNA sequence encoding a complete variable region in the germline. The actual V-D-J joint region specifies a major binding site (largest of three such sites) in the fully folded antibody protein which makes direct contact with the antigen molecule.

There are two exceptions to the general rule that no complete V gene exists in germline DNA. In skates (cartilaginous fish) one L chain gene family in the germline consists of 'fused' VJ sequences.[6] Either this family represents a primordial gene that predated the separation of V- and J-elements, or it arose by joining of V- and J-elements. The latter interpretation is favoured by the group of Gary Litman, discoverers of these genes. The second exception is a group of heavy chain chicken V pseudogenes which also contain 'D-bits' fused to the end of the V sequence, all in the *preferred* reading frame used in functional VDJ rearrangements.[7] One possible way these genes could have arisen is by reverse transcription of V(D)J mRNA in a B cell producing cDNA retrotranscripts with an inframe VJ or VDJ joints which somehow became integrated in the germline.

However, what is of most relevance to the main message of this book is that knowledge of the detailed molecular differences between the 'germline configuration' and the 'somatic configuration' has made it possible for us to obtain and interpret molecular data in terms consistent with a Lamarckian soma-to-germline feedback process in the immune system. The contrasting germline versus somatic gene configuration is one of the main focal points of discussion in the following chapters, as it defines the genetic uniqueness of the immune system and is at the core of our assertion that the antibody (and TCR) genetic systems constitute what might be called in lay terms 'smart genes'. The unique features of Ig genes, and the characteristic use of diagnostic molecular patterns produced during their evolution in the vertebrates, provide a basis for a new interpretative framework for understanding the new genetics of antigen-recognition molecules. It is based on the known RNA splicing patterns in V(D)J messenger RNA, reverse transcription (copying of RNA back into DNA), and proposed mobility of DNA sequences from the soma to the germline.

For those readers thinking ahead this statement may at first be intellectually jarring from a number of points of view. One apparent paradox generated by such an outcome is: if evolution, in all its 'wisdom', generated a genetic strategy capable of producing more than a million different antibody specificities to enable vertebrate responses to the unexpected, then why would such an efficient system ever need to superimpose a Lamarckian V gene feedback loop? The answer may be that the extant system may not have emerged without somatic hypermutation and soma-to-germline feedback. We will return to address these issues later in the book.

A DIGRESSION: CAN WE COMPARE THE IMMUNE SYSTEM WITH CURRENT COMPUTER SOFTWARE?

There is a computer-related analogy that seems appropriate at this point. Infectious electronic 'viruses' invented by our cyberpunk cult are but one of the many hazards we now face in the age of the Internet connecting millions of personal computers around the globe. Many appear almost biological in their behaviour because when they infect hard disks, they can erase or corrupt both software and files, and they can make more copies of themselves to ensure their own survival as we unwittingly send off infected messages to colleagues and friends. Anti-virus software packages are now available which automatically scan incoming files for known computer viruses. But what about entirely new electronic viruses? How can we protect our computers against them? How does current anti-computer-virus technology measure up against the biological strategy of the immune response? What if cyberpunks were able to create new electronic viruses with all the efficiencies and inbuilt strategies of our own immune system? Could the Internet or major telephone exchanges and switching hubs begin to suffer from an AIDS-related illness? Would this have to occur before computer programmers were stimulated to develop anti-viral software that is able to respond to the unexpected in a more organic way?

These are intriguing questions. IBM are currently creating an 'immune system' to fight viruses in cyberspace.[8] At present anti-virus programs live in a host computer to monitor system functions, program changes or family signatures in order to detect and combat a virus. However, in IBM's digital immune system currently being developed, any data sample thought to be infected is automatically sent to a virus analysis machine. A 'decoy' program lures the virus to infect it so that the viral code is brought out of hiding for further analysis. The virus analysis machine then updates its own files before sending this new information back to the infected machine—plus any other potentially infected node in the network (immune response). Thus, the new anti-viral programming techniques are being inspired by elementary models of the vertebrate immune system. 'Evolution' is already in action in the digital battleground. Further progress in anti-virus technology may continue to undergo parallel changes with our growing understanding of biological evolution.

In this chapter we have described many examples of how investigating the efficiency and logic of the immune system may help reveal some fundamental principles for those interested in 'organic' computer program design. Although there are good reasons why our airplanes do not resemble birds, these discussions also illustrate some important lessons. Birds have much more highly evolved stability and manoeuvrability factors than an airplane. Yet, our planes can fly faster. Nature tends to be flexible and small, while our own designs are larger, heavier and more rigid. For years, we have used nature as the inspiration for our technological designs. Today, as our understanding of nature, both physical and biological, is approaching smaller and smaller units for analysis, an entirely new class of intelligent systems can be conceived. In computer security (cryptography), we see computer programmers starting to invent security systems based on fundamentally new quantum physics, such as that based on qubit logic. In materials systems engineering the aim is to design more intelligent materials which can change their own shape, monitor their own health, and control vibrations. Yet in working towards the development of these biologically inspired systems, we still have a long way to go to approach the level of efficiency of the adaptive immune system. As Richard Feynman has noted, when considering the efficiency of nature at the quantum scale in physics and whether information could be efficiently processed at that level, 'There's plenty of room at the bottom . . .'[9]

In solid state and nuclear physics, scientists are only now discovering the details of some of nature's rules for the vastly different logic and physical nature of subatomic particles. Just as there is the potential to eventually produce nanocomputer technology (of the order of 10^{-9} of a metre or 10^{-3} microns), there is also the potential for computer technology to operate with the same efficiency as the living cell with a whole range of inbuilt intelligent functions housed and internally managed in an area of about 5 to 10 microns (the approximate diameter of a cell nucleus is of the order of 10^{-5} metres). In the near future, using current manufacturing techniques, advances in ultraviolet lasers, robots and lenses needed to etch microcircuits onto silicon chips could extract a new generation of microprocessors with features as small as 0.13 microns across. Beyond this there are several technical and physical barriers which make further miniaturisation improbable with existing technology. There is also the potential for computer scientists to emulate some of nature's information handling strategies to improve programming efficiency and functionality. Yet, it is also here that the popular metaphor with current anti-virus scanning devices breaks down. When we compare current anti-viral software strategies with the strategies of our own immune system it seems we have much to learn. Information technology will have to make quantum leaps to approach the sophistication of the vertebrate immune system which has been shaped over evolutionary time scales.

CHAPTER 5
SOMATIC MUTATION

In Chapter 4 we raised the possibility that somatic mutations of V genes may have been an essential part of the evolution of the adaptive immune system as we know it in extant higher vertebrates. Here we will begin the discussion with a series of rhetorical questions and general answers.

First: 'If we could "play God" with our genetic material (DNA) how would we generate a *non-random* pattern of mutations in a DNA base sequence?' The only scientifically logical answer is: 'Given that mutations occur by *random* errors in the DNA sequence' there is only one way, by imposing a selection criterion on either the mutated DNA sequence or on its protein (or RNA)[1] product so that those mutated genes that successfully meet the selection criterion are retained, and those that do not are discarded.'

Second: 'How do we decide that enough mutation has taken place so that we do not overdo the mutational change by allowing a successful, selected mutant to change to an unsuccessful sequence?' Again the scientifically logical answer is: 'By designing a feedback signal in the test for successful mutated genes to ensure that they do not mutate further.'

Finally: 'How do we ensure that only one gene mutates?' (as most genes must be conserved to ensure

125

proper function). And the answer is: 'By using a unique molecular shape as a docking device so that the mutation machinery docks (binds) exclusively to the gene locus targeted for mutation.'[2]

These are challenging issues if viewed through the dogmatic prism of contemporary neo-Darwinism. Indeed the choice of words is inherently contradictory. Classical genetics asserts that almost all gene mutations are harmful. This traditional view now needs qualification because of a special class of gene mutations in the immune system of vertebrates which confers a selective advantage on the animal. These mutations are manifest in the antibody variable genes V(D)Js expressed in mature B lymphocytes and are found in high affinity antibodies (which fit better to foreign antigens) produced after an immune response has been in progress beyond about a week (Table 5.1). Such antibodies are characteristic of long-lived memory B lymphocytes. Somatic mutation thus allows the immune response to generate better fitting antibodies with the potential to more efficiently clear infecting agents from the body. Thus, the 'play God' scenario outlined above has actually evolved over millions of years of vertebrate evolution, via the biological process of 'trial and error' interactions between molecules.

Some 25 years ago the mere mention of the words 'somatic mutation' amongst immunologists was guaranteed to generate heated and lively discussion. In the intervening years the debate has cooled and given way to a vast amount of molecular evidence documenting the reality of high-rate antigen-driven somatic mutation, now dubbed 'somatic hypermutation', targeted to antibody variable region, V(D)J, genes.

It is interesting that in recent experiments done by the group led by Nobel laureate Rolf Zinkernagel,[3] it has been shown that the immune response against certain viral proteins does not display affinity maturation based

126

Table 5.1 Some basic terms relevant to somatic hypermutation and the soma-to-germline feedback loop (also see Glossary)

Affinity maturation Mutant antibodies produced by the memory B cells which emerge from a Germinal Centre are of higher binding affinity for the antigen than the antibodies produced during the early phase of the immune response.

Alleles/polymorphism Alternative DNA sequences of the same gene which change the structure of the protein slightly are called *alleles*. A population of individuals expressing different alleles of the same gene is said to be *polymorphic* for that gene.

Apoptosis The term used to describe a process of biologically *controlled* cell death.

Germinal Centre A specialised area of a lymph gland or spleen which displays intense cell proliferation (by cell division) and cell death (by *apoptosis*). Most of the cells in a Germinal Centre are B lymphocytes. Germinal Centres are caused by antigenic stimulation and they are the site of *somatic hypermutation* and antigen-mediated *affinity maturation*.

Germline configuration Refers to the *germline variable (V-) element* which *has not* been rearranged to J or D/J elements.

Histones In higher cells the very long DNA strands of the chromosomes are bound to histone proteins in ordered loops which allow packaging of the DNA into the small space of the nucleus.

Hybridoma A normal B cell can be 'immortalised' by a technique of cell fusion with a continuously dividing B cell tumour cell to create a so-called 'hybridoma'. It can be grown indefinitely in tissue culture and it secretes antibody of only one specificity (that of the original B cell).

Locus-specific device A term used to describe a docking (binding) site for the RT-mutatorsome which allows the focusing of somatic hypermutation to *only* the V(D)J and its contiguous immediate flanking DNA sequence.

mRNA Messenger RNA which is transcribed (copied) from the DNA base sequence of a gene.

Mutagen Hazardous chemical or penetrating radiation which can cause mutations.

Point mutation Refers to a single base change. For example, a nucleotide

127

Table 5.1 (continued)

base (**G**) changing to any of the other 3 bases (**T**, **A** or **C**, see Figure 2.5).

Rearranged variable region gene   Also referred to as *V(D)J*, which can apply to a rearranged H (VDJ) or L (VJ) chain variable gene.

Replicase/polymerase   A generic term for a polymerase enzyme that copies DNA or RNA base sequences into progeny DNA or RNA base sequences. DNA polymerase (DNA replication) and RNA polymerase (transcription) and reverse transcriptase (copying an RNA template sequence into DNA) can all be classed as 'replicases'.

Reverse transcriptase   An enzyme which copies an RNA base sequence into a DNA base sequence.

RNA intermediate   Refers to the fact that a copy of a gene (DNA sequence) made into RNA (by transcription) can be reverse-transcribed into cDNA or a retrotranscript.

RT-mutatorsome   The hypothesised molecular organelle within a B cell responsible for somatic hypermutation (RT = reverse transcriptase). See Figure 5.6.

Soma-to-germline feedback loop   Refers to a process whereby *somatically mutated V(D)J* genes are reverse-transcribed to *cDNA* which then recombines with the most similar (homologous) germline V gene sequence so as to replace it.

Somatic configuration   Refers to the *rearranged variable (V) gene* which is *only* found in mature B and T lymphocytes. Generic symbol is V(D)J.

Somatic hypermutation   Refers to the mutation process activated by antigenic stimulation of a B cell targeted to *rearranged variable region genes V(D)Js* of mature B cells in *Germinal Centres*.

Template molecule   Any single-stranded DNA or RNA sequence which serves as the template for copying by a 'replicase'.

---

upon somatic hypermutation. Furthermore, high affinity antibodies are no better at clearing viral infection than lower affinity antibodies. The conventional argument is that somatic hypermutation is needed for high affinity antibodies that confer improved protection against infection. If the Zinkernagel group's results are confirmed more widely with other infections, in other extant vertebrates, it would imply that optimally functional antibodies could be selected from the *pre-existing* repertoire without the need for mutation during the course of infection. That is, in an evolutionary context, somatic hypermutation and affinity maturation may now be redundant.

However, at the beginning of the evolutionary development of the adaptive immune system, when the germline V gene repertoire was very much smaller than in modern vertebrates (it must have started with a single gene), it is obvious that somatic hypermutation would have conferred a great selective advantage. This advantage would take the form of a greatly expanded repertoire of antibodies expressed in B lymphocytes during the life of an individual animal. Further, if the soma-to-germline feedback loop is true, then somatic mutation of antibody variable region V(D)J genes could be a critical evolutionary tool to accelerate the building of germline V gene repertoires. Available information is compatible with this scenario. Somatic hypermutation *is present* in all jawed vertebrates which possess an adaptive immune system, including the most primitive examples, the cartilaginous fish.[4]

The phenomenon of somatic hypermutation has been demonstrated most clearly in mouse B lymphocytes; an analogous process occurs in human B lymphocytes. Hypermutation in higher vertebrates also operates, albeit slightly differently, in B lymphocytes of chicken, rabbit and sheep.[5] The manipulative experiments addressing the mechanism have been conducted mainly in genetically modified ('transgenic') mice particularly from the Cambridge University laboratory of Nobel laureate Cesar Milstein and his collaborator Michael Neuberger.[6] This technology has not yet been applied in other species.

In this chapter we will first discuss how gene mutations in general are produced and then concentrate the discussion on the phenomena (and historical development) of the process of somatic hypermutation in V(D)J genes. We will argue that this process is highly regulated, involving a 'locus-specific device' which focuses mutations to the DNA sequence of the somatically rearranged V(D)J coding region, thus protecting the constant (C) region sequences and all other genes in the genome from the deleterious effects of random mutation.

GENE MUTATIONS ARISE FROM ERROR-PRONE COPYING INVOLVING RNA INTERMEDIATES

A change in the DNA base sequence, as we recounted in Chapter 2, is defined as a 'mutation' (Figure 2.6). If the mutation—the base change—occurs in the region of the DNA that codes for protein, it will change the triplet codon and potentially change the amino acid specified by that codon. Thus, a different amino acid inserted in the protein chain can result from a single base change in the DNA sequence (termed a 'point mutation'). Most mutant proteins do not function normally (although rarely they may perform an entirely different function in another physiological or metabolic context). Therefore most mutations are deleterious, conferring a selective disadvantage in terms of conventional, Darwinian 'survival of the fittest' cell or multicellular organism.

Figure 5.1 illustrates this idea for a normal and a mutant protein. Notice that a single point mutation of a $G{\rightarrow}C$ in the codon normally specifying the amino acid 'aspartic acid' (Asp) generates a new codon specifying the amino acid 'histidine' (His) at that position in the protein chain (and see Appendix). Such point mutations can have radical consequences in that the mutant protein could have

130

Figure 5.1 Diagram showing how point mutations can generate new mutant proteins with a different folding pattern

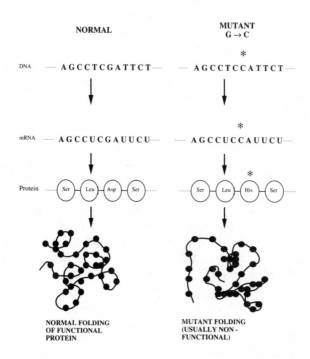

Note: For additional background information see Tables 1.1, 5.1 and Figures 2.5, 2.6 and 2.7.

a different folding pattern and therefore a potentially different function (i.e. normal function would have been lost). This principle was graphically demonstrated 40 years ago when the amino acid sequence of the β (beta) chain of human haemoglobin was determined both from healthy individuals and those suffering from sickle-cell anaemia, a hereditary disease in which the oxygen-carrying haemoglobin molecule is defective. A haemoglobin protein molecule is a heterodimer consisting of two α (alpha) chains, or α subunits, and two β chains or β subunits. The genetically determined disease is manifest in so-called 'homozygous' individuals where both of the diploid chromosomes carry the defective β chain gene; 'heterozygotes' who carry one normal copy of the β chain gene and one mutant copy are carriers of the gene and display a mild condition called 'sickle-cell trait' (as half of their haemoglobin molecules within a red cell would have normal oxygen-carrying capacity). It was found that β chain proteins isolated from people with sickle-cell anaemia had the amino acid 'valine' instead of 'glutamic acid' at position 6 in the 146 amino acid long chain. This single change was shown to be responsible for the formation of large aggregates of mutant haemoglobin molecules within the red blood cell, deforming the cell and causing it to take the shape of a sickle. Interestingly, the sickle cell gene persisted in Africa because heterozygotes had a selective advantage when infected with malaria, as the parasites cannot replicate as efficiently in a sickle cell as in a normal red blood cell.

This again brings us back to a key question: how do gene mutations arise? The traditional view, until just recently, was that they occurred by chance and quite mysteriously. They were said to arise 'spontaneously' and the usual causative culprits are the cosmic rays from the sun and other types of electromagnetic radiation (X-rays, ultraviolet light). Charles Darwin called many of them 'sports' and assumed their random generation by the conditions of 'Nature'. While chance indeed plays a role—different mutations can have different probabilities of occurring—there is now, following 30–40 years of accumulated research in virology and molecular biology, far less mystery surrounding how they might arise. The late Darryl Reanney, formerly of La Trobe University, Melbourne, published seminal contributions to rationalising this area in the early 1980s.[7] Through his analyses, and the earlier work of Nobel laureates Arthur Kornberg, Manfred Eigen and Howard Temin (who discovered reverse transcription in RNA tumour viruses) and other virologists such as John Holland, we now have a coherent way of viewing the mechanisms by which many gene mutations may arise. It is all about the fidelity of synthesis of DNA or RNA sequences copied from either DNA or RNA template molecules, in processes involving the four types of nucleic acid-copying enzymes: DNA polymerase, RNA polymerase, RNA replicase and Reverse Transcriptase.

Studies at the molecular level have shown that the suite of enzymes that replicate the hereditary DNA (i.e. the chromosomes) have several error-correcting and editing functions.[8] Mutations such as base substitutions generated during DNA replication are low probability events (Figure 5.2). The maximal error rate is probably less than one base change in 100 million bases replicated ($<10^{-8}$) although the true error rate is probably much lower at $<10^{-10}$ (less than one in 10 billion bases replicated). This is extraordinarily high fidelity information copying. As the DNA polymerase, the 'DNA copying machine', moves along the template DNA sequence it constantly checks the new complementary copy for errors. It can do this because the new strand of DNA immediately starts complementary base pairing to produce the duplex DNA typical of the double helix. Aberrant base pairs such as **T** pairing with **G**, or **C** pairing with **A** are instantly detected as distortions in the shape of the right-handed helix. The DNA polymerase enzyme complex then clips out the incorrect base (or group of bases) and inserts the correctly pairing bases (**A** or **G** respectively). The ability to rapidly check the new strand by its base pairing within the double-stranded DNA helix is the key to understanding how the editing

Figure 5.2 Diagram showing error rates of DNA and RNA synthesis

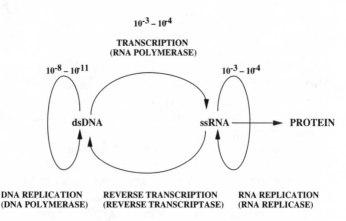

$10^{-3} - 10^{-4}$

**TRANSCRIPTION
(RNA POLYMERASE)**

$10^{-8} - 10^{-11}$ $10^{-3} - 10^{-4}$

dsDNA   ssRNA ⟶ PROTEIN

| **DNA REPLICATION** | **REVERSE TRANSCRIPTION** | **RNA REPLICATION** |
| (DNA POLYMERASE) | (REVERSE TRANSCRIPTASE) | (RNA REPLICASE) |

Note: The error rates refer to incorrect bases incorporated per base per copying event (see also Figure 2.4); ds = double-stranded, ss = single-stranded.

and repair functions of the DNA polymerase-complex evolved. The speed of accurate replication in bacteria is about 500 bases per second, and in higher cells (including human cells) it is about 50 bases per second. The DNA sequences of higher cell chromosomes are far longer and the chromosomes more complex than the smaller and simpler bacterial genomes. In higher cells, unlike bacteria, the DNA in the chromosomes is complexed with proteins called 'histones' which are concerned with folding the long DNA strands into a series of loops to enable them to be packed into the nucleus. DNA replication is initiated concurrently at *many sites* on each

134

chromosome, thus ensuring that the large set of DNA sequences is replicated in 5–20 hours.

Recall that in Chapter 2 we discussed the high error rates for the production of RNA from a DNA template (transcription) and for the production of DNA from an RNA template (reverse transcription). Both of these error-prone gene copying processes generate point mutations at a rate of one in 1,000 to one in 10,000 bases replicated (10^{-3} to 10^{-4}), rates which are vastly higher than the error rate in DNA replication of chromosomes (10^{-8} to 10^{-10}). RNA viruses, such as influenza virus, replicate their genomes with a comparably high error rate, which explains the rapid genetic change leading to global flu epidemics. HIV alternates error-prone copying between RNA→DNA (at the integration step) and DNA→RNA (at the expression step during an infectious cycle) and it also displays a very high mutation rate. Thus, all copying involving single-stranded RNA intermediates (to turn RNA into DNA and vice versa) are 'sloppy' or low-fidelity gene copying processes. Repair of the sequence cannot be effected because this requires the unaltered template strand to engage in double-stranded duplex formation with the new strand to allow template-directed correction of the error. Further experiments have shown that the enzymes involved in error-prone polynucleotide copying (RNA polymerase, Reverse Transcriptase and RNA replicase) do not possess proof-reading and error-correcting repair functions.

The knowledge summarised above raises the possibility that a proportion of mutations appearing in DNA have not appeared in a haphazard, mysterious fashion, but have arisen through natural error-prone copying via the low-fidelity copying loops involving RNA intermediates depicted in Figure 5.2.

Genetic mutations of the normal type that are passed on to offspring occur at low frequency. These are 'germline' mutations. They are rare, successful mutants that have escaped the efficient proof-reading functions of the DNA polymerase at the time the germline DNA was replicated and packaged into the male and female gametes (sperms, eggs). As we can observe and characterise their defects (i.e. their effect on the 'phenotype' and health status of the individual), we can determine that they are being propagated at low frequency in the population. As is the case with the 'sickle-cell' phenomenon, such mutations are most destructive in homozygotes (where both gene copies are defective) and are moderated in the heterozygote because the protein product of the good gene copy can compensate partially for the mutant protein.

However, there are variants of particular genes (alternative forms called 'alleles') expressed in different individuals that are not necessarily life-threatening or disadvantageous. These germline variants constitute the normal variability in populations of organisms that Darwin assumed existed before natural selection acted. An important question is: how have these benign alleles arisen?[9] Conventional neo-Darwinian thinking asserts that alleles all arose by chance mutation in germline DNA and have been established in the population of organisms (the so-called 'gene pool') by natural selection. Later (pp. 198–205) we examine whether alternative explanations are possible in the context of soma-to-germline feedback.

While it is not proved, there are suspicions nowadays that increases in congenital abnormalities and spontaneous abortions have been caused by environmental factors such as toxic chemicals polluting the environment in which we live. For example, it is now common knowledge that the towns and cities surrounding heavily polluted inland seas such as the Aral Sea in the former Soviet Union suffer from a shocking elevation in birth abnormalities. Similar claims have been made by Vietnam veterans and the North Vietnamese as a result of exposure to chemicals like Agent Orange, a toxic defoliant used during the Vietnam War. Chemicals that act on genes to change the DNA coding sequence are called 'mutagens' and may operate by overloading the normal DNA repair processes. Phenomena called 'error-prone DNA repair' have been identified in bacterial cells and in higher cells overwhelmed by DNA damage. As in the case of alleles, the deleterious effects of mutagens are conventionally thought to operate on germline DNA.

GENERAL IDEA OF SOMATIC MUTATION

A multicellular organism consists of hundreds of millions of cells, some of which are being constantly replaced at extremely high rates. For example, in humans and other vertebrates all cells in the blood—white and red—are turning over at a rate of tens of millions each day. The epithelial cells of skin and mucosal surfaces (gastrointestinal, nasal and pharyngeal cells) constantly divide, producing millions of new daughter cells each day to replace effete cells that are shed from the epithelial surface. In contrast, lower rates of cell turnover occur in the internal organs such as heart, liver, kidney and brain. Neurones (nerve cells) do not divide at all in adult humans. When a cell divides to produce daughter cells, the DNA in the nucleus replicates to produce a copy of all the chromosomes for delivery to the two daughter cells. When large cell numbers are involved we must expect some 'somatic mutations' to occur (even though the error rates of mutation during DNA replication are low). We therefore expect that in large multicellular animals somatic mutations will be appearing all the time, particularly in those cell populations or tissues where the cell turnover is very high.

All cancers are caused by somatic gene mutations: a deleterious change of the genetic constitution of the cell that grew into the cancer during the life of the host animal. The cancer cells may have abnormal function and be unresponsive to normal growth-limiting or death-inducing signals, taking on 'a life of their own'. They may mutate further, become locally invasive or give rise to satellites (metastases) like an autonomous organism within the body of the host. Obvious examples of somatic mutations are the skin cancers. These may arise from a single mutant cell by cell division. A clone of cells—like a localised colony of fungal cells growing as a mould on stale bread—grows at a local site on the skin. Some of these cancers are pigmented (melanoma) and can be highly malignant. They are now known to be caused by exposure to ultraviolet radiation from the sun.

Somatic mutations leading to cancer, like those adversely affecting the function of essential structural or enzymic proteins, are obviously detrimental. We have discussed them briefly here to provide a contrast with the beneficial mutations that are known to occur in the antibody variable region genes and to emphasise the extraordinary nature of the control/selection processes in mutating B lymphocytes. We will now reiterate the key issues and explore the possible biological mechanisms involved.

It is now established that somatic mutations in re-arranged antibody variable region genes, V(D)Js, can arise at high rate during the course of an immune response. An antigen-selected B cell can mutate its V(D)J genes at a rate of about 1/1,000 to 1/10,000 bases per replication event (expressed as 10^{-3} to 10^{-4}). This rate is of the order of a million times faster than normal background mutation rates in other, germline-transmitted, genes. The appearance of mutant antibody V-regions occurs within five to ten days following antigenic exposure. The phenomenon of increased antibody affinity for antigen (see Figure 3.8) is now thought to be based on ongoing somatic mutation

138

and selection during the course of the immune response. The molecular mechanism of the mutation process is under active investigation in a number of laboratories including those of Ted Steele and Bob Blanden.

There are still key issues to be explored. Can a biological system ensure that somatic mutations will be beneficial (for example, by removing deleterious mutations, while saving only those which are useful)? As we will soon explain, the answers to this question is yes. The immune system has worked out a tightly coupled somatic mutation/selection-of-the-fittest strategy which confers a potential benefit on the animal.

FINE STRUCTURE OF VARIABLE REGIONS—'WU-KABAT STRUCTURES'

Before we address the molecular and cellular details of somatic mutation it is first of all important to describe the fine structure of antibody variable regions which make up the antigen-binding site of the antibody (TCR structures of T lymphocytes are similar in principle).

Almost 30 years ago Elvin Kabat and associates began to define the amino acid sequence structure of antibody variable regions. As the protein sequences of many human antibodies began to accumulate, T.T. Wu and E. Kabat could line up all the available heavy and light chain sequences and construct a graph of amino acid variability for each position in the protein chain of the antibody variable region. An example of this type of data presentation, which we call a 'Wu-Kabat plot', is shown in Figure 5.3. The plot of amino acid variability at each position in a population of antibodies reveals three regions of hypervariability and other segments called the framework regions (FRs). The hypervariable regions form a surface which makes intimate contact with the surface contours of the antigen. The framework

139

Figure 5.3 Wu-Kabat plots showing that hypervariable stretches in the variable regions of H and L chains of antibodies coincide with antigen-binding sites called 'complementarity-determining regions'

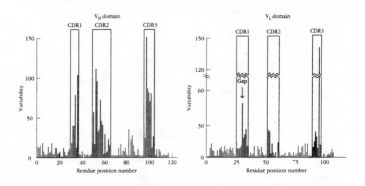

For a large collection of human antibodies of different specificity, relative variability can be calculated at each of the 100 or so amino acid positions along the length of the variable regions of the heavy and light protein chains. These frequency plots show that variability is *not* uniform along the length of the variable region—there are three clear regions of hypervariability (HV) bracketed by regions of low variability (sequence conservation) called the framework regions (FW). The HV regions are known to coincide with those regions of the protein chain which make direct contact with the antigenic epitopes—such regions are called complementarity-determining regions or CDRs. The numbering starts at the tips of the molecule (see Figure 3.2) so that CDR3 is adjacent to the constant region. Variability patterns such as these can be called 'Wu-Kabat structures'—the highly non-random patterns indicate that the variable sequences have undergone positive Darwinian selection by antigen binding at the level of the protein HL heterodimers. V gene families from all vertebrate species display the Wu-Kabat variability pattern. See Tables 3.1 and 5.1 for further information. (From J. Kuby *Immunology* 3rd edition W.H. Freeman & Co., 1997. Published with permission of publisher. Based on E.A. Kabat et al. 1977 *Sequence of Immunoglobulin Chains*, US Department of Health Education and Welfare.)

regions provide the basic structural 'skeleton' or backbone of the V-regions, allowing the formation of the antigen-binding surface. Depending upon the nature of the selecting antigen, the antibody surface may vary from a relatively flat surface with small indentations or projections, to a 'pocket' which receives a projection on the antigen surface. High affinity antibodies have surfaces with good complementarity (fit) to the antigen surface.

The original Wu-Kabat plot can therefore be thought of as the amino acid distribution of 'successful' antibody variable regions that could bind to a variety of different antigens. In more recent work the different antibodies used to construct a Wu-Kabat plot have been somatically mutated variants of a germline V sequence driven by a single antigen. The Wu-Kabat plot clearly shows that variations appearing in antibody variable regions are non-random.

Indeed, if antibody variable region V(D)J amino acid sequences in a data set give a plot like that in Figure 5.3 (what we will call a 'Wu-Kabat structure'), it is strong presumptive evidence for antigen-binding selection of the HL antibody protein heterodimers in that data set. In short, a 'Wu-Kabat structure' (peaks of sequence variation in the regions of antigen contact) is an indication that antigen-binding selection has fashioned the amino acid sequence variability of the population of V(D)J sequences under study. We will return to this important point in the next chapter in the context of attempting to explain why 'Wu-Kabat structures' appear in populations of unrearranged germline V-elements.

It will be recalled from Figure 4.5 that a complete variable region gene is a somatic construct of V_H, D and J-elements in the H chain and V_L and J-elements in the L chain brought together by DNA rearrangement in a B cell. Two regions of hypervariability are located in the germline encoded V-element, but a fully assembled third

141

hypervariable region, which spans the joint of a V-D-J or a V-J rearrangement, is *never found* in the germline of higher vertebrates although there are exceptions in cartilaginous fish (skate, horned shark) and chicken V pseudogenes. The third hypervariable region is generally essential for antigen recognition. An antibody that lacks it (containing only the two germline-encoded hypervariable regions) could not bind to an antigen. Therefore, the germline V genes cannot be subjected to direct, conventional Darwinian natural selection. The selection process must be indirect and must involve consequences of V-D-J and V-J rearrangement which only takes place in the soma.

THE 'GERMINAL CENTRE': SOMATIC HYPERMUTATION IN REARRANGED V(D)J GENES

All available evidence indicates that in B lymphocytes only the rearranged V(D)J genes that encode the antibody protein produced by that B cell somatically mutate. In other words, variable genes remaining in the germline configuration do *not* accumulate mutations in somatically mutating B lymphocytes. Therefore the first level of control, allowing the emergence of 'beneficial mutations' in the immune system, involves DNA rearrangement to provide a target V(D)J sequence upon which the mutator acts. The next level of control involves the mutational machinery itself, which must bind to a unique structure associated with the rearranged V(D)J sequence ensuring that only this part of the DNA sequence on the chromosome mutates.

The other important point to note here is that the somatic hypermutation process in mice and humans is *caused by* antigenic stimulation. In other species (such as sheep) somatic mutation may be triggered by other signals (as yet undefined) in specialised gut-associated lymphoid tissues. Thus, variability in lymphocyte V(D)J genes is caused by an identifiable environmental influence or specialised signal. This underlines the point that during the somatic evolution of the immune system, genetic variability can be *caused* by the environment. This contrasts with the neo-Darwinian dogma that all variability in germline genes pre-exists *before* the natural selection forces act. A critical experiment proving this important point was performed by Ursula Storb and associates in the mid-1980s. They introduced a limited number of copies of a known rearranged light chain (VJ) gene into the germline DNA of an inbred mouse strain (the animals are called 'transgenic' mice). They showed that if these VJ-transgenic mice were left unimmunised the VJ-transgene did not mutate. They could sequence the piece of DNA containing the VJ-transgene from the B lymphocytes of the spleen of the transgenic mice and show it was unmutated. However, upon immunisation with the antigen known to bind antibodies expressing that particular VJ variable region sequence, the target VJ-transgene accumulated numerous somatic mutations.

Where does somatic mutation take place? Can it occur in the mobile B cells in the blood and lymph or does it occur in specialised locations? In the mid-1980s David Gray and Ian MacLennan working at the University of Birmingham predicted that somatic mutation took place in a specialised post-antigenic location called a 'Germinal Centre'. Such tissue sites had been known for many years to be regions of intense B lymphocyte proliferation in lymph nodes and spleen following antigenic stimulation. It was also known that 80 per cent or more of B cells in the Germinal Centre appeared to die there. In the 1960s Gus Nossal and Gordon Ada working at the Walter and Eliza Hall Institute for Medical Research in Melbourne established by tracer studies, using radioactively labelled protein antigens, that undegraded forms of the antigen complexed with anti-bodies localised on the surface membrane of cells in the Germinal Centre (and could stay there for many months). The antigen–antibody complexes were located on the surface of large specialised cells that could spread projections of their cell membrane (called 'dendritic projections') throughout the Germinal Centre, forming a network. These cells are known as 'follicular dendritic cells' (FDCs).

Follicular dendritic cells interface with the B cell populations of the Germinal Centre (Figure 5.4). Gray and MacLennan predicted that somatic mutations were being generated in the rapidly dividing B cells of the Germinal Centre. Those 'successful' progeny cell mutants that could bind and compete for the antigen molecules displayed in the antigen–antibody complexes on the FDCs would survive and all other mutants would die. This was indeed a bold theory of affinity maturation! It coupled rapid somatic mutation with an antigen-binding selection process to rescue from death those mutants which successfully competed for antigen on FDCs, by virtue of the high affinity of the new mutant antibody.

Within five years, groups led by Garnett Kelsoe, Klaus Rajewsky and Claudia Berek proved that the first part of the Gray-MacLennan prediction was correct by showing that targeted somatic mutation of rearranged variable V(D)J genes was indeed going on within B cells of the Germinal Centre. Kelsoe and Rajewsky went a step further and showed that dividing B cells which were not in the Germinal Centre did not mutate their V(D)J regions, indicating that the critical signals activating the 'mutator' were within the Germinal Centre. At about the same time, Ian MacLennan and his group established that if the presumed progeny B cells bearing somatically mutated Ig molecules on their surface were treated in such a way that the membrane Ig was cross-linked (a

Figure 5.4 Somatic mutation and selection of mutant antibody-producing cells in a Germinal Centre

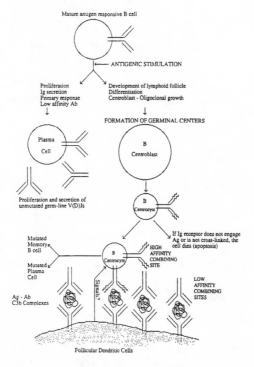

Following 'selection' of a circulating B cell by antigen, the activated cell will either secrete the antibody at high rate or go on to act as a founder cell for a Germinal Centre, where it will mutate its rearranged variable region V(D)J genes. The antibodies produced early in the response form antigen–antibody complexes which bind to the surface of the Follicular Dendritic Cells (FDCs). These antigen-presenting cells form an extensive membranous network throughout the Germinal Centre. After a rapid cell proliferation phase (B centroblasts) smaller non-dividing B centrocytes test their mutant antibodies for binding to the antigens displayed on the FDCs. If the mutant antibody is non-functional or of low binding affinity the cell will die in a programmed way called 'apoptosis'. If the mutant has a *higher* binding affinity than the antibody complexed to the antigen on the surface of the FDCs, it will dislodge it and bind the antigen. This leads to a signal being delivered to the B cell,

145

preventing it from dying by apoptosis. The selected mutant B cell migrates out of the Germinal Centre and either becomes a plasma cell (producing and secreting high affinity antibodies) or a long-lived memory B cell. Somatic mutants of a given V(D)J amino acid sequence, when lined up as in Figure 5.3, will generate a 'Wu-Kabat structure' indicative of the fact that the antibody variable regions have been selected by antigen binding. (Republished with the permission of the publisher Munksgaard International Publishers, Copenhagen, from Steele, E.J. et al. *Immunological Reviews* vol 135: 5–49,1993.)

Note: Ab, antibody; Ag, antigen; C3b, activated blood proteins which bind to Ab–Ag complexes.

process mimicking antigen-binding), such cells would *not* die.

Thus by 1991, we had a picture of the Germinal Centre as a lymphoid tissue site of active somatic hypermutation, a regulated process coupled to antigenic selection. We want to emphasise that the Germinal Centre in both man and mouse is a structure which develops only during the immune response to a foreign antigen. We have summarised the main events in Figure 5.4.

A simple reconstruction of the time–space sequence of Germinal Center genesis, based on experimental evidence, is as follows. A B cell in the circulation is selected by binding with foreign antigen. It migrates to the white blood cell areas of a spleen or lymph node. After receiving a special signal from a T helper cell, it starts to proliferate. These activated B cells produce antibodies which form antigen–antibody complexes that localise on the surface of follicular dendritic cells (FDCs). Some of the descendants of the original selected B cell seed special lymphoid sites containing FDCs called primary follicles. One or more of these selected founder B cells rapidly proliferates (called a 'B centroblast' in Figure 5.4) and no longer has an antibody receptor on its surface membrane. This proliferation phase (cell division every five to seven hours over about five days) leads to a population of up to 20,000 daughter cells which are smaller (called 'centrocytes'), no longer divide and again display antibody receptors on the cell surface membrane. At some stage in the production of the centrocyte population, the rearranged variable V(D)J genes hypermutate. Now the Germinal Centre is mature, a complex post-antigenic structure containing B centrocytes, helper T cells and FDCs. The latter cells form an extensive network displaying antigen–antibody complexes involving the antibodies produced in the first days of the response (Figure 5.4).

The following events then appear to take place. Firstly, the thousands of centrocyte cells constitute a huge repertoire of antibody-displaying cells, with their surface antibodies encoded by somatic mutants. Most of these antibodies (about 80 per cent or more) no longer bind the antigen. As in other proteins, most mutations result in a change in the shape of the antibody which prevents it from fitting the shape of the antigen. Rare mutations result in antibodies with a better fit (higher affinity) for the antigen than the original antibody. The new antibodies are positioned on the cell surface of B centrocytes and can compete for antigen molecules that are arrayed in the antigen–antibody complexes on follicular dendritic cell surfaces. However, to compete successfully with the antibody in the complexes (produced in the first day or two of the response), the new mutant antibody must be of the *same or higher* binding affinity. This is the essence of the affinity maturation mechanism: competitive antigen-binding selection. The Germinal Centre is a transient 'mutant breeder organ' for V(D)J genes and only the fittest B cells survive. The unsuccessful (low affinity, nonfunctional) B cell mutants (the majority) die in a process of programmed cell death known as 'apoptosis'.

To summarise, B cell clonal selection occurs at two levels. First, a B lymphocyte is selected by the foreign antigen from a diverse population of cells. That B cell centroblast proliferates rapidly leading to a population of up to 20,000 new somatic variable region V(D)J mutant B cells. Second, mutant, high affinity B cell lines are antigen-selected, survive, proliferate, and secrete high affinity antibodies or become long-lived memory cells. There is *nothing* mystical about this scenario. It is conventional Darwinian natural selection operating on a large, variable cell population from which only a minority survive, those with antibody mutants of the highest binding affinity for the antigen. The Germinal Centre is therefore a cauldron of genetic hypervariability and selection in the immune system. But it is tightly regulated and controlled. The B lymphocytes mutate their V(D)J genes (and only these genes, no others) within the highly specialised and regulated environment of a Germinal Centre.

So far we have emphasised the role of conventional Darwinian natural selection at the cellular level in the immune system. However, we have also introduced a *distinctly* neo-Lamarckian concept, namely a role for an environmental signal (the antigen) in the generation of the genetic diversity upon which natural selection operates. Could this radical evolutionary implication have stimulated part of the scientific animosity against the idea of 'antigen-generated diversity' first proposed by Alistair Cunningham in the mid-1970s?

VALIDATION OF THE FACT OF ANTIGEN-DRIVEN SOMATIC MUTATION

About 25 years ago Alistair Cunningham, working at the John Curtin School of Medical Research in Canberra, first articulated the concept of *antigen-driven* somatic mutation in antibody variable genes.[10] He also provided the first experiments demonstrating the phenomenon. At about the same time, the research groups of Melvin Cohn

148

(Salk Institute, San Diego, USA) and Susumu Tonegawa (Basle Institute for Immunology, Switzerland) provided the molecular proof (from protein and DNA sequencing) that the hypothesised process of somatic mutation was a demonstrable experimental fact.[11] Both of these break-throughs were taking place at the height of the somatic versus germline debate in the early 1970s. The earliest theory of the molecular mechanism of somatic mutation, provided by Sydney Brenner and Cesar Milstein (Cambridge University, England), involved a type of error-prone DNA replication. The normal high fidelity copying of a DNA template to produce another DNA sequence was proposed to be deficient in the case of mutation of antibody variable region genes (the deficiency being due to the inability of the DNA polymerase to correct its errors when replicating the DNA encoding antibody variable regions). Thus the earliest models of somatic mutation were DNA- and cell division-based; as antigen-activated B cells rapidly proliferated, variable region mutants appeared, targeted to the DNA encoding the variable region.

Like most novel scientific hypotheses, experiments and interpretations, the pioneering work of Alistair Cunningham was controversial. However, by 1981 a spate of experimental data from the laboratories of Patricia Gearhart, Ursula Storb, Al Bothwell, Leroy Hood, David Baltimore and Klaus Rajewsky proved the correctness of Cunningham's central prediction, i.e. mutations appearing at high frequency in the DNA sequence of rearranged variable region V(D)J genes occurred *after* 'antigenic stimulation' in mice. Further proof was provided within a few years by a new generation of experiments involving transgenic mice.

A major technical advance which made all of this work possible involved the 'hybridoma' or 'monoclonal antibody' technique developed in 1975 by George Köhler

149

and Cesar Milstein (1983 Nobel Prize shared with Nils Jerne). The technique utilised fusion of a tumour cell (which can grow and divide indefinitely in culture) with an individual B cell from an immunised animal. The hybrid cell produced the antibody of the B cell and acquired the immortality of the tumour cell. Such hybridomas allow the experimenter to demonstrate the B cell clones predicted by Burnet's Clonal Selection Theory (page 97), with all cellular descendants having the same rearranged V(D)J genes and thus making the same antibody. Such a clone of hybridoma cells enabled the accurate comparison of the DNA sequence in the somatically rearranged B cell V(D)J gene with the germ-line sequence of the V-element. It could be shown in many cases that B cells isolated *after* antigenic stimulation had accumulated many somatic mutations in the region from amino acid residue (or codon) position 1 to about position 85 in the V-element sequence, which takes in the first two hypervariable regions (see Figure 5.3). This is consistent with the idea of antigen selection in Germinal Centres. The mutation rates were very high. For example, in a region of DNA around the rearranged V(D)J gene, experimenters could find that the DNA sequence had accumulated changes that differ from the unmutated germline sequence in about 5 per cent of the bases.

HOW ARE THE MUTATIONS DISTRIBUTED IN THE TARGET REGION?

Exactly where in the rearranged B cell DNA do the somatic mutations occur? We now know that they are localised to a short stretch of DNA sequence. This experimental evidence is summarised in Figure 5.5, where we show a rearranged heavy chain VDJ gene at the start of the Germinal Centre reaction and the same sequence

Figure 5.5 Diagram showing that the rearranged variable region V(D)J gene and adjacent flanking region DNA is the target of the mutator

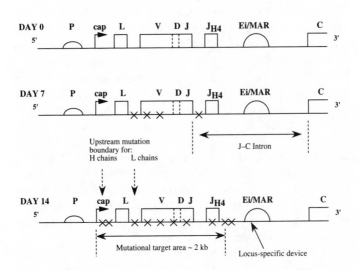

This is an example for human or mouse heavy chains. The pattern of mutations for the major family of light chains (termed *kappa* chains) in human and mouse is very similar. The diagram illustrates that V(D)J sequences isolated later in the immune response (day 15 after antigen) display a higher incidence of somatic mutation. Note that the somatic point mutations are distributed from the vicinity of the start site for mRNA synthesis (transcription) to a region in the J–C intron near the Ei/MAR region. Note also that the promoter region (P) further upstream of the VDJ and the downstream constant (C) region are not mutated. Ei/MAR (J–C intronic Enhancer/Matrix Attachment Region) is proposed to be the 'locus-specific device' on which mutator machinery docks; X = Point mutation; Cap site = start site of mRNA synthesis; kb = kilobases, 2kb = 2000 bases. The protein coding regions are shown as rectangular boxes. For additional information see Table 5.1 and Figures 2.6, 4.5 and 5.1.

seven and fourteen days later. Notice that the incidence of mutation appears to increase with time after antigenic

151

stimulation and that mutations are confined to the immediate environment of the rearranged VDJ and the
non-coding flanking DNA sequence. They are not found
further upstream, near the Promoter site (P), where the
RNA polymerase enzyme complex binds and initiates the
copying of the mRNA from the DNA template (at the
'cap' site indicated by the small curved arrow just
upstream of the Leader 'L' coding region). Nor are they
found to any significant extent downstream beyond what
we predict is the 'locus-specific device' (the Ei/MAR, the
intronic Enhancer/Matrix Attachment Region).

Within the VDJ coding region itself, the mutations
found in a collection of functional high affinity antibody
sequences are distributed in a non-random fashion, that
is, they display a 'Wu-Kabat structure'. Thus, point
mutations within codons which result in a changed amino
acid (called 'replacement' changes) tend to accumulate
in the hypervariable regions (or those parts of the protein
which form the antigen-binding site). Other point mutations which are 'silent' (those changes within a codon,
see Appendix, that do not result in a change in the
amino acid) accumulate in the conserved framework
regions. These Wu-Kabat structures (page 140), as
indicated earlier, mean that the somatic mutant antibodies
were subjected to antigen-binding selection.

This targeted distribution illustrates two key principles of somatic hypermutation. The separation of DNA
sequences encoding variable (V) and constant (C)
regions implies the evolutionary advantage of allowing
mutational change in the V genes whilst rigorously
conserving the C genes. The focusing of the somatic
mutator mechanism to the rearranged variable V(D)J
region and not the C-region gene is particularly important for heavy chains. This is because the heavy chain
genes encode that part of the antibody molecule which
determines functional properties such as: promotion of

lysis of bacterial cells; the uptake and destruction of infectious agents by phagocytes; signalling the B cell for proliferation and antibody production; key processes leading to the stimulation of T cell help for B cells. This focusing is not needed in the case of light chains because light chain constant regions do not have these functions.

The second principle illustrated by Figure 5.5 is that the mutations do not extend upstream of the gene. The 5′ boundary is in the vicinity of the transcription start site or the Leader intron (the non-coding intervening sequence between L and V coding regions, see Figure 4.5). There is a good reason for this because this upstream region contains key regulatory sequences (the Promoter or P region) which allows the binding of the RNA polymerase-complex initiating transcription and the production of the mRNA (at the position indicated by the directional arrow or 'cap' at the 5′ end of the mRNA).

The essential point is this: the mutations are precisely focused so as not to mutate the C regions and not to mutate the upstream P regions which control the expression of the gene. How is this the critical targeting achieved?

MECHANISM OF SOMATIC HYPERMUTATION OF V(D)J GENES

The distribution of mutations shown in Figure 5.5 plus the known error rates of nucleic acid copying involving RNA intermediates (Figure 5.2) were two key facts which led Ted Steele and Jeff Pollard in 1987 to propose the prototype 'Reverse Transcriptase Model' of somatic hypermutation (RT-model for short).[12] The idea began life in Wollongong in February the year before and was completed by the northern summer of 1986 when Ted met with Jeff in New York City. (At this point they thought they understood most of its implications and decided to publish the idea.) Since the time of its conception Ted has always considered the RT-model to be inspired by, if not *derived* from, the Somatic Selection soma-to-germline idea. However, subsequent critical arguments by Bob Blanden has convinced Ted that the RT-model of somatic hypermutation must be logically *prior* to the Somatic Selection theory (and must also be prior in the evolutionary sense—somatic mutation of a small starting set of germline V genes would logically precede soma-to-germline flow of V gene information, see pp. 183–5). This epistemological twist actually simplified the interpretation of DNA sequence data, particularly those aspects of the genetic recombination signature of germline V genes (see the discussion of the soma-to-germline 'integration footprint' pp. 180–2 and Figure 6.3).

Since 1986, further work to investigate the Reverse Transcriptase hypothesis has continued in our laboratory involving colleagues Gerry Both and Harry Rothenfluh. We can now outline a detailed theoretical molecular model of somatic hypermutation within a B cell: it involves error-prone reverse transcription and gene feedback to DNA (summarised in Figure 5.6). The model is supported by the vast majority of experimental results concerning somatic hypermutation. It can be extended to explain molecular mechanisms which operate to somatically diversify chicken rearranged variable region V(D)J genes, hitherto termed 'gene conversion'.[13] However, we must emphasise that until all of the molecular detail has been experimentally validated, what follows is theory, albeit theory consistent with *all* the available evidence.

We have proposed that the molecular machine which mutates rearranged V(D)J DNA at a high rate be termed the 'RT-mutatorsome' (RT = reverse transcriptase). There are a number of molecular organelles with the

154

'-some' suffix such as the 'ribosome' (the protein–RNA complex required for translation of the messenger RNA into an amino acid sequence, see Appendix) and the 'spliceosome' (also an RNA–protein complex which splices out introns from the pre-mRNA). Thus, the proposed RT-mutatorsome uses the unspliced pre-mRNA as a template for synthesising cDNA. The term cDNA, where 'c' stands for complementary, is the generic term for all copies of an RNA template converted into a DNA sequence by reverse transcription (cDNAs are also called 'reverse transcripts' or 'retrotranscripts').

We have proposed that the reverse transcription process which produces the mutated cDNA copy of the rearranged variable region V(D)J target area is initiated in a special region or 'primer' site downstream of the V(D)J near the Ei/MAR region (Figure 5.6) and proceeds from right to left, to the cap site (the 5′ end of the pre-mRNA template). Cesar Milstein and colleagues have shown experimentally in transgenic mice that the 'locus-specific device', Ei/MAR, is essential for somatic hypermutation, while the V(D)J coding region and the Promoter can be replaced by haemoglobin gene counterparts without detriment to mutation. We propose Ei/MAR as the 'locus-specific device' needed to dock the RT-mutatorsome to the rearranged V(D)J gene and focus mutation to the V(D)J. (We are currently experimentally testing this idea.) We also propose that the mutated cDNA copy of the V(D)J region is integrated back into the chromosome and displaces the original, unmutated V(D)J (indicated by the looping arrow in the figure). This type of genetic integration has been demonstrated experimentally in many organisms and is called 'homologous recombination', as it involves the alignment of very similar DNA sequences followed by recombination of DNA strands. This theoretical scenario ensures that

155

Figure 5.6 Schematic diagram of the reverse transcriptase (RT)-mutatorsome

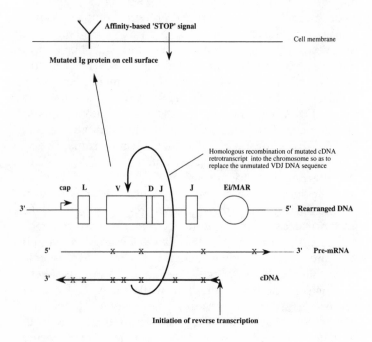

This is an example for human and mouse heavy chains. The RT-mutatorsome is expected to act in a similar fashion for the major family of light chains (termed *kappa* chains) in human and mouse. Many of the copying and recombination events depicted in the figure take place in the nucleus of the mutating B cell. The 'locus-specific device' (Ei/MAR) docks the RT-mutatorsome ensuring that reverse transcription of the pre-mRNA template is initiated upstream of the Ei/MAR but downstream of the target V(D)J. This is accomplished because all nucleic acid (or polynucleotide synthesis) is *always* in the 5′ to 3′ direction, which means that all templates which are copied have to be in the anti-parallel 3′ to 5′ orientation. The DNA template strand for RNA synthesis is the 3′ to 5′ DNA strand. At the

156

cap site, pre-mRNA synthesis is initiated and since it is error-prone there is a significant chance each transcribed copy of the V(D)J region could carry base substitutions (indicated as X). Reverse transcription, which is also error-prone, is initiated on the pre-mRNA upstream of Ei/MAR and proceeds back towards the 5′ end of the pre-mRNA. Mutated retrotranscripts (cDNAs) are homologously integrated as indicated by the looped arrow so as to replace the original unmutated V(D)J sequence. Mutated RNA transcripts are then processed (e.g. introns spliced out) and exported to the cytoplasm where they are translated into H and L chains to allow formation of the antibody protein (Ig) which is tested at the surface of the B cell for binding to antigen presented by FDCs. Successful high affinity binding of antigen provides a 'mutation STOP' signal. See Table 5.1 and Figures 4.5, 5.4 and 5.5 for additional information about symbols and concepts. (Adapted from Steele, E.J., Rothenfluh, H.S. and Blanden, R.V. *Immunology and Cell Biology* vol 75: 82–95, 1997.)

upstream Promoters and downstream Constant regions are both protected, but the non-coding DNA in the immediate environs of the V(D)J is mutated at a very high rate. (The *same* rate of error generation known for transcription and reverse transcription as indicated in Figure 5.2, of about 10^{-3} to 10^{-4} per base per copying cycle.)

Thus the molecular directional rules of copying from DNA or RNA templates coupled with intrinsically error-prone RNA and cDNA synthesis are entirely consistent with the requirements for somatic mutation. On the pre-mRNA, there is only one direction in which cDNA copies can be made and that is backward to the start site for transcription (the cap site). If cDNA synthesis were initiated at or near the Ei/MAR region this would automatically ensure targeting of the V(D)J for mutation without the risk of mutating Promoter and Constant regions: to immortalise the mutant sequence within the body, homologous recombination is then required to integrate the mutated cDNA copy into the chromosomal DNA, thus ensuring transmission to descendant daughter cells.

In the mouse, the 5′ or upstream boundary for mutation is near the cap site for H chains and somewhere in the L-V intron for light chains (Figure 5.5). These sites conform with the two main places where cDNA synthesis will terminate, either: (a) when the reverse transcriptase comes to the 5′ end of the pre-mRNA template (the cap); or (b) around the L-V intron, because the intron may be removed in the splicing process that converts pre-mRNA to mRNA.

SOMATIC MUTATION DATA OUTSIDE THE AMBIT OF CONVENTIONAL DNA-BASED MODELS YET PREDICTED BY THE REVERSE TRANSCRIPTASE (RT)-MODEL

What other evidence demarcates the RT-model from all those mutation theories we have classed as 'DNA-based' (i.e. which depend on localised error-prone DNA synthesis in the vicinity of the rearranged V(D)J region)? The first point is a general negation: there is no evidence of a mechanism which selectively stops DNA synthesis. Once begun, synthesis proceeds until the end of the template is reached. Thus, special rules of DNA synthesis would need to be invented and experimentally validated to sustain such theories.

The other type of data concerns two sets of findings (from the laboratories of Patricia Gearhart and Eric Selsing) which are inconsistent with DNA-based models but which are predicted by the RT-model. In one type of experiment a 'reporter' sequence was located just downstream of the rearranged V(D)J region (between VDJ and J in Figure 5.6). This reporter sequence had the effect of suppressing mutation in the V(D)J target area, a result incompatible with DNA-based theories, but consistent with the RT-model. The interpretation is as follows. The reporter sequence (a short piece of DNA encoding a special type of RNA involved in protein synthesis called 'transfer RNA' or 'tRNA') is such that when transcribed into RNA it folds into the tRNA structure, a three-dimensional shape that fits to and transports specific amino acids. In the RT-model, because the reverse transcriptase would not proceed through the tRNA structure, cDNA synthesis would terminate before reaching the V(D)J. Two other properties of tRNA could also terminate reverse transcription. First, the tRNA sequence could be cleaved and released from the pre-mRNA sequence, thus prematurely ending the RNA template on which the RT-mutatorsome copies RNA into cDNA. Second, known chemical modification of the nucleotide bases in the tRNA could inhibit cDNA synthesis by the RT-mutatorsome.

The other type of experiment concerns homologous recombination. It was shown in a carefully engineered transgenic mouse, using two slightly different rearranged variable V(D)J genes linked close together, that somatic point mutation of the target V(D)J *always involved* homologous recombination with the other closely related V(D)J sequence. The *obligatory association* of somatic point mutation with homologous recombination is the striking finding. This result is nonsensical in terms of DNA-based mutator theories because in these theories, homologous recombination plays no role. However, as can be seen in Figure 5.6, the RT-model demands 'homologous recombination' as integral to the mutator process. That is, point mutations do not accrue in the target chromosomal DNA sequence unless the mutated cDNA homologously recombines so as to displace the original, unmutated V(D)J sequence.

Therefore, at this point in time the reverse transcriptase-based mutator theory provides the best expla- nation for all the present evidence on somatic

159

hypermutation. The process hypothesised over ten years ago has not been refuted by any of the numerous experiments performed since. It has passed what the late philosopher of science Sir Karl Popper would call some 'severe tests' in the scientific arena.

WHAT IS THE SIGNAL FOR SOMATIC MUTATION TO STOP?

At the start of this chapter in the 'play God' scenario we asked: 'How do we decide that enough mutation has taken place so that we do not overdo the mutational change?' We would argue that it is a selective advantage (in a traditional Darwinian sense) that when an antibody has mutated to a point where it possesses high affinity for the antigen in the Germinal Centre, a signal should be given to the B cell to stop it mutating further. In this way, mutation is not overdone and the risk of losing binding affinity for the antigen is minimised. We have therefore also predicted that the signalling function of the surface-bound Ig receptor has been selected during evolution. When a mutant B cell successfully binds the antigen presented on the follicular dendritic cell (FDC) in the form of antigen–antibody complexes, a 'STOP' signal is delivered to turn off the somatic mutation process (Figure 5.6) by turning off production of the protein subunits of the RT-mutatorsome. (Proteins are continuously degraded by enzymes generically termed proteases. They therefore disappear unless continuously replaced by new protein synthesis.)

This point allows us to deal with another central issue which has had a tacit but profound grip on mainstream thinking about somatic mutation. All DNA-based models of mutation are cell division-dependent. The essential assumption is that DNA replication (and thus cell division) must take place for mutations to be generated (by

160

a localised error-prone DNA repair process). Indeed the current mainstream model is that mutations are generated during the rapid cell division phase (in centroblasts) which provides the large B cell population (centrocytes) in the Germinal Centre (Figure 5.4). However, the RT-model is not dependent on either DNA replication or cell division—mutations can be generated in the absence of either process. DNA replication is only required to propagate the mutated V(D)J sequence in dividing B cells. Cell division would be useful in B cells with high affinity antibody, but wasteful in mutant B cells which had lost affinity for the antigen.

The paradigm that 'cell division is necessary for mutation' has become so deeply entrenched in the thinking of many scientists that the rate of somatic mutation is written as '10^{-3} to 10^{-4} per base per cell generation'. In other words, the cell division concept is embedded within the conventional description of mutation rate and tacitly biases thinking about the mechanism.

Because the RT-model is cell division-independent, we favour a cyclic process based on 'mutate→pause (for expression of Ig)→affinity test' occurring within a non-dividing cell which is transcriptionally active, i.e. producing mRNA and protein molecules of H and L antibody chains and the components of the RT-mutator-some. This is a far more efficient way of organising the mutation/selection process so as to maximise the production of high affinity antibodies than cell division-based models.

'DIRECTED MUTATION' AND INHERITANCE OF SOMATIC MUTATIONS?

We have described here our proposal for a reverse transcriptase-based, error-prone, V(D)J gene feedback

161

process within a mutating B lymphocyte. All current experimental evidence is consistent with this theory. This, we have argued, is at the heart of antigen-driven somatic mutation of antibody genes. It leads us to a second distinctly neo-Lamarckian concept: the idea of a 'directional' gene feedback loop.

Rapid random mutation involving the replacement of the original DNA sequence coupled to selection for a better one (encoding an antigen-binding site with higher affinity than the original antibody) effectively results in the appearance of 'directed mutation', the complete antithesis of random mutation.

We are now ready to ask the big questions. Given the fact of extensive somatic mutation during immune responses, does *any* of this somatic genetic variability play a role in evolution? More specifically, we can ask: is there any evidence that acquired somatic mutations in variable region genes of antigen-selected antibodies can contribute to the next generation? Put another way, can acquired somatic mutations be inherited in the germline DNA? Can the homologous recombination event that we have proposed to take place within an individual B lymphocyte extend to the transfer of a V coding sequence from a B lymphocyte to the germline DNA in sperm or eggs?

CHAPTER 6
SOMA-TO-GERMLINE FEEDBACK

This chapter will outline our explanation of how the germline DNA sequences of antibody variable genes have changed in response to antigenic stimulation over evolutionary time.[1] The evidence points to the operation of a gene feedback loop involving the flow of genetic information from somatic cells (lymphocytes) into the germline genes. We begin with a question first posed by Ted Steele some twenty years ago[2] in relation to the state of knowledge at that time on the evolutionary genetics of antibodies: 'How do we reconcile the inheritance in the germline DNA of a characteristic which by an established set of criteria only arises in the soma?'

The same type of question has been nurtured over the years by a select group of philosophers and biologists troubled by conventional explanations of the mechanism of evolution. Since the publication of Darwin's theory it has manifested itself, in various guises: in the writings of the nineteenth-century philosopher Herbert Spencer, the writer Samuel Butler, the writer and philosopher Arthur Koestler and the zoologist Frederic Wood Jones.

In the past, neo-Darwinians have dismissed the implications of this question as irrelevant because Weismann's doctrine specifically forbids the inheritance of acquired characters—propositions of this type are therefore not true. Closed-loop tautologies like this

163

abound in the contemporary scientific literature to the
point that some of our most prominent science writers
in biology such as Stephen Jay Gould can make the
categorical statement:

> Nature, . . . works on Darwinian, not Lamarckian, prin-
> ciples. Acquired characters are not inherited, and desired
> improvement occurs by rigorous selection with elimination
> of the vast majority from the reproductive stream.[3]

Indeed in the same book Gould makes no mention
of Darwin's 'Pangenesis' theory nor even of August
Weismann's famous 'soma–germ plasm' tissue barrier.
The issue has therefore been marginalised as the 'inco-
herent rantings of madmen' (read 'Lamarckians', prefixed
with a 'neo' or otherwise). However, to be fair to Gould,
he is very clear in what domain Lamarckian inheritance
can operate:

> But cultural change, on a radical other hand, is potentially
> Lamarckian in basic mechanism. Any cultural knowledge
> acquired in one generation can be directly passed to the
> next by what we call, in a most noble word, education
> . . . This uniquely and distinctively Lamarckian style of
> human cultural inheritance gives our technological history
> a directional and cumulative character that no natural
> Darwinian evolution can possess.[4]

In the early 1950s F. Wood Jones in his book *Trends
of Life* wondered just how long this tragic misunderstand-
ing could persist. Arthur Koestler made eloquent pleas
in this regard in the 1960s and 1970s. In his last full
book,[5] published in 1978, *Janus: a Summing Up*, he
distilled the inherent contradictions of the neo-Darwinian
position and argued that the door be opened just a little
to admit the legitimacy of Lamarckian explanations in

biology. For a rational understanding of structure versus function relationships in biology, he considered it almost nonsensical that evolutionary geneticists had erected a 'genetic chastity belt' around the reproductive organs. Why could we not admit that the Weismann Barrier could be *selectively permeable* to somatic genetic information over evolutionary time?

Today the question has gained a special kind of urgency and poignancy because of the precision that molecular biology has brought to our understanding of the structure and function of genes. We predict that the paradoxical appearance in the germline of a gene structure thought to be present only in the soma will be hard to ignore early next century as molecular biology finalises the DNA sequencing of the human genome (and others). The vast mass of new DNA sequence data will require innovative, open-minded interpretation. Lamarckian possibilities will have to be admitted into global scientific interpretation in biology. We anticipate that the molecular signature of somatic genetic processes 'written into' germline genes will become difficult to avoid and will need to be confronted at both the scientific, clinical and wider sociological levels.

Indeed this may not be the first time mankind has had to deal with the resurrection of an idea previously marginalised by mainstream thinking. In another era and involving another syndrome, the varied psychological pathologies amongst shell-shocked survivors of World War I provided Sigmund Freud with much overt public evidence for the existence of what he called the 'unconscious'.

For the variable genes of the immune system we now restate the question (first posed in 1992): 'How do we explain the presence of highly non-random patterns in the germline V gene DNA which can only arise by direct antigen-binding selection forces acting on the protein product of the gene (the antibody) and *not* the DNA directly?'

Given that 'Wu-Kabat structures' can only be fashioned during evolution at the level of antigen-binding by functional protein heterodimers (the complementary folding of the variable region portions of H and L protein chains making up an antigen-combining site), how is it possible for non-expressed germline V-gene segments to show highly detailed evidence of antigen-binding selection? Providing a rational answer to this question is the driving force behind our current research program.

INHERITANCE OF SOMATIC MUTATIONS?

Therefore we can question whether the DNA sequences of germline V genes benefit in any way from the antigen-driven somatic mutation/selection events that have occurred in previous generations. We described in the previous chapter how *many* new V-region mutant DNA sequences appear in B lymphocytes during immune responses and are selected by the success of the encoded antibody in competing for antigen. We now ask if these new sequences can spill over into the germline DNA of the germ cells, eggs and sperm.

The 'Somatic Selection Theory' predicts the germline transmission of acquired somatic mutations of antibody V-region genes. It could be effected via the agency of the enzyme reverse transcriptase (copying somatic RNA into DNA) plus the ubiquitous, naturally occurring, endogenous RNA retroviruses (produced by lymphocytes) acting as 'gene shuttles' ferrying mutated V-region gene sequences into germ cells. This would then be followed by the physical integration of this somatically derived genetic information into the germline DNA so as to replace a pre-existing gene sequence (for an outline see Figure 1.2). At the time of its formulation it was

166

stressed that this scenario was a 'useful heuristic device' like that used in physical theory. In this case it allows the scientist to focus rational attention onto the possibility of the inheritance of acquired characters. It persuasively suggests that since Weismann's Barrier can be easily penetrated theoretically, why not in practice?

At the time of its publication, *Somatic Selection and Adaptive Evolution* was received with mixed reactions ranging from violent criticism, through cautious evaluation to generous praise. Some critics simply dismissed it as fanciful at best as it required scientists to embrace Lamarckian heresy. Professor Jan Klein, the editor of *Immunogenetics,* panned the book, stating: '. . . Unless he [Steele] is willing to admit that genes, cells, organs, organisms—or God—all know what they are doing (and thus make his view truly Lamarckian), he is in the same boat as the Darwinists—only his boat is leaking.'

Others were more generous, particularly the philosopher Sir Karl Popper, who thought it the most exciting book he had read that year, and Sir Peter Medawar, who had 'no idea what the outcome would be but I hope Steele is right'. Richard Dawkins, in his book *The Extended Phenotype,* considered the whole episode a 'Lamarckian scare' going so far as to state:

> I use the word 'scare' because, to be painfully honest, I can think of few things that would more devastate my world view than a demonstrated need to return to the theory of evolution that is traditionally attributed to Lamarck. It is one of the few contingencies for which I might offer to eat my hat.[6]

We will return to Dawkins' critique of the somatic selection hypothesis later in this book.

Some critics of the somatic selection idea have considered it too complex, as it requires the sequential occurrence of *too many* separate, if not improbable, events, namely appearance of the mutation in somatic DNA/RNA, its cellular selection by antigen, its transport to the germ cells, the copying of RNA into DNA (the copied product is called cDNA if it arises by reverse transcription), and then the physical, genetic recombination event leading to the integration of the mutated somatic cDNA copy into the germline DNA (Figure 1.2). However, at the time of the formulation of this theory, it was well known and accepted that B lymphocytes stimulated by contact with antigen produce copious quantities of endogenous retroviruses, supposedly harmless viruses carrying the enzyme reverse transcriptase. We still argue that the Somatic Selection Theory provides one plausible scenario based on the evolutionary advantage of feeding somatically mutated V sequences back into the germline; endogenous retroviruses can be considered sophisticated gene-packaging vectors mediating genetic communication between cells.

At the rhetorical level, therefore, the main justification of the Somatic Selection Theory is that it allows critical evaluation and interpretation of whole vistas of molecular facts which would have remained unrecognised. This is the same as the argument in favour of the 'Punctuated Equilibria' concept used recently by Stephen Jay Gould[7]— the soma-to-germline feedback idea is useful not least because it has brought the supposed impermeability of Weismann's Barrier into the rational scientific spotlight.

In fact, in 1994, in the prestigious *Proceedings of the National Academy of Sciences* of the United States of America, Simona Bartl, David Baltimore and Irving Weissman speculated that genes could be transmitted between species by viral infection, thus contributing to the evolution of the vertebrate immune system. Implicit in this speculation is the penetration of August Weismann's barrier. It is even more demanding than Ted

168

Steele's idea, which does not require viruses to move from one animal to another. Interestingly, we are not aware that the scientific community and the neo-Darwinists have reacted to this speculation in the way that they did to the Somatic Selection Theory. Perhaps it is because Lamarck was not mentioned. Perhaps it is because David Baltimore is a Nobel laureate and he and Irving Weissman are members of the Academy. Whatever the reasons, it is an interesting example of the sociology of the scientific community.

PATERNAL TRANSMISSION EXPERIMENTS

Initial and highly controversial experiments[8] addressing this question were conducted in the late 1970s, not at the level of the genes but at the level of functional aspects of an immune response. Ted Steele asked whether altering the immunological function of a parent (the father) by active treatment with antigen led to any alteration in specific or non-specific immune reactivity in the progeny resulting from mating such an antigen-treated father with a normal female. These experiments were conducted in inbred strains of mice and were controversial because of inconsistent results—some laboratories could demonstrate paternal transmission of specific acquired, somatic immunological function (or non-specific abnormalities) and other laboratories could not. The first positive demonstration was reported by Ted Steele in collaboration with Reg Gorczynski using Medawar's system of acquired neo-natal tolerance to foreign transplantation antigens (see Figure 4.3). These experiments were conducted in late 1978 and into 1979 at the Ontario Cancer Institute in Toronto. We showed that if we repeatedly exposed newborn male mice of strain **A** to large numbers of lymphocytes of strain **B**, these 'tolerant' males could transmit aspects of this

169

specific transplantation tolerance (to **B** tissue antigens) to their progeny on mating to normal females of strain **A** (Figure 6.1). The immune effector cells in these experiments (whose function was 'suppressed') were almost certainly T lymphocytes, as they are involved in graft rejection responses. In later experiments, antibody responses of B lymphocytes were examined, showing the paternal transmission of alterations in the magnitude of the antibody response in progeny of immunised males.[9]

The problems with these experiments were logistical and technical. They involved breeding programs of a long-term nature encompassing ten to twenty treated and untreated male parents and the screening of 50 to 100 animals at any one time for alterations in the response of progeny animals to skin grafting or immunisation, or the antigen-stimulated response of their lymphocytes in test-tube assays. Adequate large numbers of control animals and reaction assays at every step were crucial for the proper interpretation of results, but created logistical and technical stress. The positive transmission results were inconsistent, occurring at a high rate in only one or two males out of ten (i.e. there was variability amongst the breeding males in terms of transmission frequency to progeny). The data suggested the direct penetration of Weismann's Barrier by active immunisation, because the male is thought to transmit only sperm in the production of the next generation. The experiments were not done with immunised mothers because the foetus and the newborn are influenced directly by antibodies and lymphocytes trafficking from the mother to the offspring, either in utero or after birth via the early milk.

Not surprisingly these experiments engendered considerable controversy when they were first reported in the late 1970s and were part of the emotive reception of the ideas mentioned earlier. The inconsistent results caused some workers to be sceptical. The biological

Figure 6.1 Paternal transmission of acquired immunological tolerance. Simplified protocol of the Gorczynski–Steele experiments

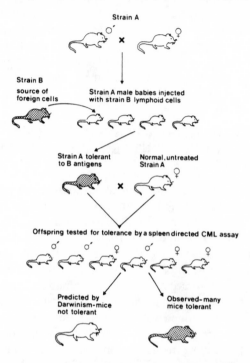

Strain A male mice are made tolerant to strain B transplantation antigens by neo-natal injection of strain B lymphoid cells. This is the Medawar strategy (see Figure 4.3). These mice are mated to normal strain A females and the progeny tested by a killer T cell assay in tissue culture for tolerance to strain B transplantation antigens. By conventional neo-Darwinian theory the progeny should not be 'tolerant', but experimentally many mice were observed to be 'tolerant'. Control breeding programs involved untreated or sham-treated males of strain A mated to normal strain A females and the parents and progeny housed under the same conditions in the animal room as the test animals. CML = cell mediated lympholysis assay for the detection of killer T cell activity. (From Steele, E.J., Gorczynski, R.M. and Pollard, J.W. in: *Evolutionary Theory: Paths into the Future*. Ed. J.W. Pollard. John Wiley, London, pp 217–237, 1984. Published with permission of the publisher.)

assays used for some of the work are inherently variable and susceptible to artefact.[10] We are now technically capable of definitive breeding experiments, because the new molecular technologies allow the production of genetically defined mice to allow the detection of soma-to-germline transmission events if they occur in real-time. However, the frequency of such events is still the big question.

THE SOMATIC MUTATION/SELECTION 'SIGNATURE' IS WRITTEN ALL OVER THE GERMLINE V GENES

Our interest in DNA sequence analysis of germline V genes began with the initial observations of our colleague Harry Rothenfluh, whose PhD studies first focused on the problem in 1992 when he was isolating and sequencing germline V gene elements from the heavy chain Ig locus in mice. The clear distinction between the 'germline configuration' versus the 'somatic configuration' of V genes has been outlined in Chapter 4, page 120. Recall that rearranged variable region V(D)J genes occur only in mature B cells and such genes are the direct targets of the antigen-driven somatic mutation and selection process in the Germinal Centre. Yet Harry Rothenfluh's data were suggesting that features of DNA sequences derived from the somatic configuration were also present in the germline configuration!

The DNA sequence structure of germline V genes poses a real problem for any contemporary evolutionary theory which relies on the Weismann prohibition of soma-to-germline flow of genetic information. First, germline V gene elements can *never* be the direct targets of natural selection forces (i.e. antigen binding). Only a fully assembled antibody protein (a H+L heterodimer) on the surface of a B lymphocyte can be selected or

172

tested for antigen-binding function. Germline V gene elements by themselves are never turned into RNA (transcribed) or protein (translated). They are only expressed in a mature B lymphocyte following DNA translocation on a somatic chromosome to give a typical variable region, or V(D)J, rearrangement (Figure 4.5). Functional studies thus far have found only about half of the human germline V gene repertoire in V(D)J sequences. Many may never be used in a mature V(D)J rearrangement in a B lymphocyte and thus may never be subjected to selection.

Second, there are numerous germline V genes in the mouse and human genome (about 100 each for the H chain and the L chain), and whilst they are very homologous in sequence within their related subgroups, they are also diverse and distinctly different in DNA sequence. The germline V genes of a homologous subset generate a 'Wu-Kabat' variability plot (see Figure 5.3), that is a highly non-random DNA sequence pattern thought only to be generated by direct antigen-binding selection acting on an intact antibody attached to the surface of the B lymphocyte. This whole issue was discussed in the previous chapter.

Third, germline V genes in mice and humans are in a relatively 'pristine' condition, with respect to the incidence of translation stop codons (see the 'Genetic Code' in the Appendix and page 109 in Chapter 4). In other words, many of the germline V genes are 'open reading frames'. There is a statistically significant deficit in the observed frequency of stop codons[11] within the V coding region from the frequency expected from a process of random point mutation. This also applies to the so-called V 'pseudogenes', non-functional V genes, which have been crippled by one or more mutations which either disrupt the reading frame of the coding regions or affect a regulatory sequence so that the gene

173

can no longer be expressed as mRNA and protein. Such crippled genes are believed conventionally to accumulate further mutations over evolutionary time, because they can never be subjected to selection. Thus, the low incidence of point mutations (or 'frame-shift' mutations due to insertions or deletions of nucleotide bases) generating stop codons in both functional genes and pseudogenes strongly suggests the operation of a mechanism that tends to maintain 'open reading frames' in the germline V gene DNA. But an open reading frame can only be selected at the level of antigen-binding, between an intact cell-bound antibody heterodimer and the antigen.

The DNA sequence structure of mammalian germline V genes therefore presents us with a profound conundrum. Random genetic drift models would suggest that such genes should accumulate their fair share of background mutations, yet this does not appear to happen—the sequences display the hallmarks of powerful and direct selection at the protein level. These features are evident in all vertebrate V gene sequence collections examined to date, from cartilaginous fish, through amphibians (frogs), to rabbit, sheep, mouse and human genomes.

This paradox deepens even further if we examine DNA sequences from germline V genes in chickens, which have been obtained over the past ten years by the French group led by Jean-Claude Weill and the American team led by Craig B. Thompson. In chickens, the process of somatic generation of antibody diversity differs from mouse and human in several important ways. First, heavy and light chain loci contain only *one* functional germline V gene element with the essential signals for DNA rearrangement by recombination-activating gene (RAG) enzymes. Upstream of this functional V gene is an array of 20–100 V pseudogenes. They are non-functional mainly because they have suffered truncations (length shorten-ings) in their transcription control regions (Promoters) which can extend into their Leader peptide (L) encoding sequences (and they lack the signal sequences for RAG enzymes). These chicken pseudogenes are close-packed in the germline DNA, with only a few hundred bases (at most) flanking each coding region. In contrast, the array of V-element germline genes in human and mouse are flanked by ten to twenty kilobases (kb) of DNA, so that the chicken IgV loci are only about one-twentieth of the size of their mouse and human counterparts.

In developing chicken B lymphocytes the single intact V gene rearranges to give a functional rearranged variable region V(D)J gene which then becomes the target for donation of variable lengths of DNA sequence (5–100 nucleotide bases) from the upstream set of V pseudo-genes. This process has been dubbed 'gene conversion'. Different gene conversion outcomes in different B cells generate a large repertoire of V(D)J sequences in separate clones of B cells. This diversity-generating strategy in chickens produces a functional repertoire of antibodies just as effective as the repertoire in mouse or humans which is generated by a multiple V gene rearrangement strategy. And the chicken requires only one-twentieth of the amount of germline DNA![12]

What is most striking about the array of V pseudogenes, however, is that when their DNA sequences are aligned using special computer programs they generate 'Wu-Kabat structures'!, see Figure 6.2. Moreover, when differences between pseudogenes are manifest as insertions and deletions of nucleotide bases in the coding regions, they are *always* in multiples of three (i.e. triplets), thus maintaining the correct translational reading frame! This is also shown in Figure 6.2. You recall that triplet codons specify the amino acids in the protein chain (see Appendix). In addition, the heavy chain

175

Figure 6.2 The DNA and amino acid variability (Wu-Kabat) profiles of germline chicken V-pseudogenes

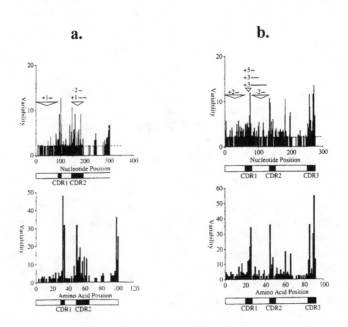

a.

b.

Wu-Kabat profiles for 18 heavy (**a.**) and 25 light (**b.**) chain variable region pseudogenes in the chicken genome. Both heavy and light chains display non-random 'Wu-Kabat structures' indicative of antigen-binding selection of B cells at the Ig protein level—a result in striking contradiction to their sequence structure which ensures that by themselves these genes are *never expressed as protein sequences*. The top profile in each set is the DNA sequence nucleotide base variability at each position across the length of the V-region and the bottom profile is the amino acid variability plot (after translating the successive DNA base sequence triplet codons into amino acids, see Appendix). The first base or amino acid of the putative coding region is assigned position 0. Note that sequence hypervariability coincides with the antigen-binding CDR regions, a pattern expected of Wu-Kabat plots of *functional variable region proteins* (see Figure 5.3). Positions where the insertion (+) or deletion (−) of base triplets (⊔⊔) occurred are indicated by open arrowheads. The relative positions of the CDR1, CDR2 and CDR3 are indicated in the diagrams below the graphs. The occurrence of base insertions or deletions *as sets of triplet codons in the correct reading frame* (see Appendix) is indicative again of antigen-mediated selection operating at the protein level—an impossibility given these genes are pseudogenes! The rational explanation is that these base sequence patterns in germline V pseudogenes are maintained as open reading frames by soma-to-germline feedback of functional V(D)J sequences after antigen-driven somatic selection. (Taken from Rothenfluh, H.S., Blanden, R.V. and Steele, E.J. *Immunogenetics* vol 42: 159–171, 1995 with the permission of the publisher, Springer-Verlag Gmbh & Co. Kg.)

pseudogenes in the chicken contain 3′ (or right-hand) ends with small sections of D sequence (all of which are in the *preferred* translational reading frame normally seen *only* in somatically expressed VDJ sequences). Both heavy and light chain pseudogenes display evidence of nucleotide base addition and trimming events known to occur somatically in B lymphocytes at the joints formed during V(D)J rearrangement which somatically creates the third hypervariable region! These are truly striking and unusual features to be found in *germline* DNA-encoding pseudogenes. They embody the 'signatures' of *somatic* events.

The chicken pseudogenes therefore pose an enormous challenge to conventional molecular genetics based on a rigid neo-Darwinian paradigm. They paradoxically display *all* the features of direct antigen-binding selection at the level of an intact antibody protein, yet their germline DNA sequences are only transcribed (into RNA) and translated (RNA information into protein) piecemeal after a *somatic* gene conversion process. Their DNA sequence structure is only rationally compatible with a genetic model which invokes a somatic V(D)J intermediate which is antigen-selected, followed by soma-to-germline feedback of the variable region sequence (most likely via RNA→DNA

Figure 6.3 Soma-to-germline feedback (for heavy chain)

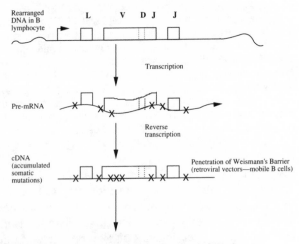

Rearranged
DNA in B
lymphocyte   L   V   D J   J

Transcription

Pre-mRNA

Reverse
transcription

cDNA
(accumulated   Penetration of Weismann's Barrier
somatic   (retroviral vectors—mobile B cells)
mutations)

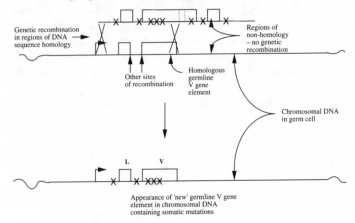

**Base pair – mediated alignment and recombination of incoming somatic cDNA sequence
with a homologous germline V-element DNA sequence**

Genetic recombination
in regions of DNA
sequence homology   Regions of
                                                non-homology
                                                – no genetic
                                                recombination

Other sites   Homologous
of recombination germline
                  V gene
                  element

                                                Chromosomal DNA
                                                in germ cell

L   V

Appearance of 'new' germline V gene
element in chromosomal DNA
containing somatic mutations

Somatic mutations accumulate in the variable V(D)J genes of a B lymphocyte
as discussed in Chapter 5 (see Figures 5.5 and 5.6). Mutated reverse
transcripts (cDNAs), delivered to germ cells by mobile B cells and/or retro-

178

viruses then undergo integration into germline DNA by the process of 'homologous recombination'. The recombination profile of germline V genes has been established by use of the phylopro genetic recombination algorithm developed by Dr Georg Weiller (see text and Weiller et al. 1998). The major sites of germline recombination are found at the transcription start site, the borders of the L–V intron and the end of the V coding region—this constitutes the 'integration footprint' discussed in the text. A less detailed version of this process is outlined in Figure 1.2. See also the annotations in diagrams and captions of Figures 4.5, 5.5 and 5.6.

copying, or 'reverse transcription') as shown in Figure 6.3 and in general outline in Figure 1.2.

The appearance of 'Wu-Kabat structures' and other somatic patterns amongst a wide range of vertebrate germline V genes and V pseudogenes, particularly the striking features of chicken pseudogenes described above, strongly suggest the operation of a somatic cell-to-germ cell gene feedback process active during the 400 to 500 million years of V gene evolution. Therefore we have solid reasons for now introducing a third distinctly neo-Lamarckian concept into contemporary evolutionary theory: the direct penetration of Weismann's Barrier in the case of some multigene families exemplified by immunoglobulin V genes.

The simplest models involve reverse transcription (such as the proposed mechanism of somatic hypermutation discussed in the previous chapter) to copy somatic pre-mRNA into cDNA followed by transport to germ tissues via mobile cells (such as lymphocytes, see pp. 185–6) or via endogenous retroviruses acting as 'gene shuttles'. The final necessary process is homologous recombination, with the cDNA replacing the original germline sequence. We have no way of measuring the frequency of such events over evolutionary time, although the strength of the 'somatic signatures' we see in the germline DNA

179

suggests that it is likely to be a frequent evolutionary event.

THE SOMA-TO-GERMLINE 'INTEGRATION FOOTPRINT'?

The Somatic Selection model to explain the evolutionary life-cycle of antibody V genes is summarised in Figures 1.2 and 6.3. Given the starting point of an array of germline V genes, we propose that new variant V sequences are 'born' in the soma by hypermutation of an antigen-selected rearranged variable region V(D)J gene expressed in a B lymphocyte. Intense cycles of somatic hypermutation and antigen-mediated selection in a Germinal Centre lead to the emergence of new, successful V(D)J DNA sequences with the mutations accumulating in the complementarity-determining regions (the CDRs or hypervariable regions which make contact with the antigen). These new DNA sequences are then transported to reproductive tissues, where they line up with a pre-existing germline V gene with a very similar sequence. A process termed 'homologous recombination' then takes place. It involves the breakage of chromosomal DNA strands and joining with the ends of the new gene to integrate it into the chromosome (this is indicated by the dotted crossed lines in Figure 6.3).

All of the features of germline V genes discussed previously in this book are consistent with this soma-to-germline 'life-cycle'. However, we have looked for additional evidence which would be consistent with a soma-to-germline flow of genetic information. We have made explicit predictions of the existence of 'an integration footprint'. If a somatic gene sequence recombined back into the germline, we would expect to see evidence for the recombination process itself (Figure 6.3). One type of evidence concerns the distribution of 'insertion/

deletion' events. When two homologous double-stranded DNA helices line up and undergo homologous recombination, the DNA strands have to be broken (or cut by a 'nuclease' enzyme) and then rejoined (by a 'ligase' enzyme). These can be imprecise processes involving nucleotide base trimmings or additions to the ends of the DNA strands, resulting respectively in base deletions or insertions at the joint. On this basis, in collections of germline V gene sequences we might expect to see concentrations of base insertions or deletions of variable length bordering the germline genes. In the same way that a 'Wu-Kabat structure' is indicative of direct antigen-binding selection, a non-random distribution of insertion/deletion events would be consistent with the operation of a soma-to-germline feedback loop. Indeed, we have found such evidence not only in our own data obtained from mice, but in *all* other relevant vertebrate germline V gene families where sufficient data are available to analyse. This non-random distribution of recombination errors fits precisely the predictions of the soma-to-germline model. At this stage in our knowledge, at least for the evolutionary genetics of germline V elements, we are not aware of alternative scientific explanations for these observed patterns (short of invoking an intelligent gene manipulator or 'divine intervener'—for which there is no evidence!).[13]

We have also recently begun to employ another predictive approach to more precisely identify recombination sites in germline V genes and thus determine more accurately what the soma-to-germline integration footprint should look like. At the time of writing we are collaborating with Georg Weiller of the Research School of Biological Sciences, at the Australian National University. Georg has developed a novel algorithm to detect recombination sites in gene sequences. Validation of the algorithm using known recombination points in bacterial

DNA sequences shows that we can precisely locate the sites of recombination in collections of germline V gene sequences. Results from this ongoing work clearly indicate that recombination in germline V genes has occurred at sites we would predict from feedback of the expected species of cDNA, the reverse transcripts arising from unspliced or spliced pre-mRNA of somatic L-V(D)J-genes.[14] We are therefore detecting in germline DNA an integration footprint which corresponds to a somatic RNA 'processing signature'! This exciting result cannot be logically interpreted using the conventional neo-Darwinian paradigm, but is entirely consistent with the predictions of the Somatic Selection Theory.

A DIGRESSION: SHOEMAKER-LEVY 9

A useful analogy to help explain our concept of the integration footprint is the size and patterns of craters on the surface of the moon and other bodies in the solar system (including sites on the earth's surface) believed to be caused by meteor and asteroid impacts. We have no doubt that such craters *have* been caused by meteor, asteroid or comet impacts at some time in the past and we can estimate the mass and dimensions of these bolides from the size of the crater. For impacts on the earth and the moon, geologists and astronomers can also estimate the time when the impacts occurred. We do not doubt this reasoning, despite the fact that we rarely witness an impact in our own lifetime. But in July 1994 we were provided with dramatic proof of the violence of such extraterrestrial impacts. We witnessed on global TV (via the *Galileo* spacecraft) telescopic real-time images of the impacts of each of the 21 fragments of the comet Shoemaker-Levy 9 as they hit Jupiter. Mankind has therefore borne witness, with a video record, to real-time events which can result in the formation of craters on the surface of planets and moons in the solar system. Sixty-five million years ago, a bolide of about 10 kilometres in diameter generated a massive crater approximately 330 kilometres in diameter near the Yucatan Peninsula off Mexico. The climatic change caused by atmospheric debris thrown up by the impact has been suggested by Luis and Walter Alvarez as a possible cause of the extinction of dinosaurs.

Yet this analogy with meteorite or comet-induced cratering is only useful up to a point. We cannot estimate exactly how often and when 'soma-to-germline genetic impacts' have occurred. But the indications from current DNA sequence data of immunoglobulin V genes are that they have occurred with sufficient frequency in the evolutionary past as to give a distinctive 'signal' of non-random sequence patterns which stand out from background random mutation 'noise'. Moreover, distinctly different non-random sequence patterns occur in the different vertebrate species, implying that soma-to-germline impact events have occurred many times *after* the divergence of these species from a common evolutionary ancestor.

EVOLUTIONARY SIGNIFICANCE OF A SOMA-TO-GERMLINE FEEDBACK LOOP?

We will now attempt to answer the question posed at the end of Chapter 4. In extant vertebrates, is there a need for a soma-to-germline feedback loop, given that their immune systems seem well adapted to handle the challenge of unexpected antigens? We will answer this question by looking at the likely events during vertebrate evolution which have resulted in the appearance of antibody genes. It is our view that the simplest logical evolutionary development of antibody V genes, consid-

183

ering all the known genetic evidence, is as detailed in the summary below.

- The first step is the emergence of a very small set of germline antigen-recognition variable genes in a vertebrate during the Cambrian explosion about 550–500 million years ago. A set of closely related though distinctly different genes can only arise by duplication of a single primordial gene followed by mutational change in the DNA sequences of the duplicates. The mechanism of gene duplication is called unequal crossing over; it occurs in sexually reproducing species during the process called meiosis which produces gametes, eggs and sperm, each with a haploid set of chromosomes. Repeated duplication followed by mutation results in a tandem (linear) array of closely related genes.[15]
- There would then be intense selection pressure on such a vertebrate to somatically hypermutate its V genes to quickly generate a larger antigen-recognition repertoire to cope with infectious diseases than that provided by the small, slowly mutating germline set.
- There would also be selection pressure to increase the germline V gene repertoire. Random germline mutation followed by natural selection would be an extremely slow way to build a repertoire. Furthermore, as we have already discussed, the evolution of heterodimeric binding sites of antibodies plus segmentation of the germline genes, necessitating somatic rearrangement of the DNA sequence before antibody can be produced, markedly blunts selection of germline V gene sequences. Furthermore, each change in germline V gene structure would require the generation of a new repertoire of germline genes. In these circumstances, soma-to-germline feedback of successful, functional,

184

mutated V gene sequences would confer great selective advantage.[16]

But in contemporary vertebrates, is somatic mutation *per se* necessary? Certainly, extensive somatic hypermutation can be demonstrated experimentally. But some experiments in inbred mice using pathogenic viruses (discussed in Chapter 3) suggest that somatic mutation either does not occur in anti-viral responses or, if it does, confers no benefit on the immune response of the mouse. Indeed, somatic hypermutation by itself appears almost irrelevant now in mouse and human. There is sufficient germline diversity and combinatorial diversity of genetic elements encoding heavy and light chains of antibodies which provides the rapid somatic generation of a large antibody repertoire sufficient to handle the unexpected. Thus, in some extant vertebrates, somatic hypermutation *per se* may be redundant. However, its role as a source of tested, successful, new open reading frames fed back to the germline may still confer selective advantage. It would act to reduce the adverse effect of random genetic drift which could potentially degrade the germline V gene repertoire by introducing translational stop codons into coding regions by point mutations or insertion/deletions of nucleotides. In short, the role of soma-to-germline feedback in contemporary vertebrates may be in 'genetic housekeeping'; it acts to maintain open reading frames in the tandem germline V gene arrays.

PENETRATION OF WEISMANN'S BARRIER

There are several ways that somatic genetic information could get into the germline. The first is via the prototypical endogenous retroviral vector. The second, a concept developed by Harry Rothenfluh, is that long-lived memory B lymphocytes migrate and penetrate into reproductive tissue and occasionally deliver new V gene sequences directly to sperm and eggs. This step could occur with or without the assistance of endogenous retroviruses although both scenarios would involve a reverse transcription step. These issues are the subjects of present and future investigation in our laboratories. But they involve complex experiments which require the collection and analysis of vast amounts of DNA sequence information. On the current scale of our operation, the outcomes may not be known for many years. However, considerable progress has been made, particularly in the past ten years, and there is no reason to think that future research will be unfruitful. Our current agenda is to operate on two research fronts simultaneously. The first is the elucidation of the mechanism of somatic hypermutation, as this process can be logically linked to a soma-to-germline feedback loop. The second, and longer-term aim, is to build genetically modified mice which will allow a definitive real-time demonstration of a soma-to-germline genetic impact. Transgenic mouse lines which express a 'kit' of a small number of human heavy and light chain V-, D-, J-, and C-region genetic elements already exist; genetic 'surgery' in the same mice (by deletion of essential DNA sequences) has ensured that the resident mouse Ig-locus genes cannot be expressed. These mice only make human antibodies when they have been immunised with antigen and undergo somatic hypermutation.[17] A molecular biologist could systematically examine the tandem arrays of *mouse* germline V genes in progeny of immunised parents for evidence of integration of human V gene sequences.

CHAPTER 7

BEYOND THE
IMMUNE SYSTEM?

A Lamarckian explanation of evolution which does not deny the validity of the Darwinian natural selection paradigm would incorporate the cause–effect sequence for the impression of an acquired character on the germline of an animal species as follows. Changed circumstances in the environment (e.g. food availability, new predators) could lead to changed habits and somatic changes in body structure and physiology. New immunological challenges could lead to new antibody genes in B lymphocytes. Over time, such changes could be integrated in the germline DNA. This would contribute to a new, enriched repertoire of hereditary variations upon which natural selection then acts to sort out the 'fittest for survival' and propagation of the line. This is the direct penetration of Weismann's Barrier. The sequence makes no statement about the cellular and molecular genetic mechanisms involved.

Charles Darwin himself held this view. According to his Pangenesis theory of use and disuse, changed environmental conditions and/or habits contributed significantly to the hereditary variation upon which natural selection acted (Figure 1.1). Why can we not easily secure direct experimental evidence to establish the truth or falsehood of such an important proposition? In the older, more measured 'neo-Lamarckian' literature, such as the work of the zoologist Professor Frederic Wood Jones, we have the following wise advice:

Those who have resorted to direct experiment have, for the most part, failed to realize the dominating influence of time in all the processes of natural evolution. Save in the production of freaks, sports or mutations, nature appears to work extremely slowly in bringing about permanent modifications in living things. This is a factor that is of necessity far more vividly realized by the paleontologist than by the geneticist . . . Moreover, the almost completely unaltered uniformity that has prevailed in certain types over vast geological periods, is a warning to the individual experimenter with the necessarily limited span of time available for experiment . . . [T]he first note of caution [is] that there is every reason to believe that, if acquired characters are inherited, their incorporation into the hereditable make-up of an animal is likely to be a slow process and, unless by reason of some unusual chance of circumstance, we would hardly look to see it accomplished in the necessarily short time available to laboratory experiments.'[1]

Emphasis therefore must be placed on the eons of time that may be necessary to effect genetic adaptations. It is likely that the successful integration of an acquired character in the germline may require repeated exposure over many generations, particularly if it is a complex character involving DNA sequence alterations to many genes. Ted Steele embarked on such breeding experiments twenty years ago with some limited and qualified success in attempts to demonstrate the paternal transmission of acquired somatic immunological alterations. He managed to show transmission through to the second breeding generation. These attempts ignored Wood Jones's wise counsel. A time-consuming logistical factor in these experiments was the need for a concurrent control breeding program with untreated parents housed under identical environmental conditions. These controls also acted as 'sentinels' reflecting unwanted changes in

188

the environment, such as mouse hepatitis virus which causes pronounced effects on immunoregulation. (This happened on several occasions necessitating abandonment of the breeding program.)

We will now discuss some 'experiments of nature' which possibly have resulted in the genetic transmission of apparently trivial acquired characters. These phenomena are useful in so far as they raise the possibility of the generality of such processes and thus stimulate speculation (and possible experimentation) on the responsible genetic mechanisms.

INHERITANCE OF CHEMICALLY INDUCED METABOLIC DISORDERS?

In the 1960s and early 1970s well controlled experiments in laboratory rats generated very interesting apparent 'acquired inheritance' or transgenerational phenomena as a consequence of experimental injury to endocrine organs. For example, disorders in blood glucose regulation involving defects in the pancreas could be induced in parental rats and be propagated to successive generations of progeny. The drug Alloxan causes experimental diabetes in rats and mice (very high blood sugar levels) by irreversibly damaging the insulin-producing cells (the β-cells) of the pancreas. Work in Japan (K. Okamoto) and later in the United States (M.G. Goldner and G. Spergel) involved Alloxan-induced diabetes in male or female (or both) parents. Diabetes developed spontaneously in large numbers of first generation offspring. Inbreeding caused the disease to become progressively worse down the generations.[2] A few years ago Ted Steele attempted to replicate this phenomenon (in principle) in inbred mice but only went to the first generation with limited success. Diabetes was induced in the male parent using the β-cell targeting drug Streptozocin. In the first generation progeny significant diabetogenic effects on early

189

development (such as increases or decreases in neonatal and weaning mean body weights) were noted although only one overt spontaneously diabetic mouse was detected.[3]

Other experiments by J.L. Bakke and co-workers involved similar types of studies on thyroid function in rats. Spontaneous metabolic disorders in thyroid function appeared in the untreated progeny of male or female parents whose thyroid function had been chemically impaired. The rate of appearance of the hormonal disorders was high, with most offspring in the litter being affected.

These very real and striking phenomena challenge us to provide explanatory mechanisms. Whilst it could be argued that the inherited effects are a consequence of the chemicals acting directly to mutate the germline genes, the possibility exists that mutations in somatic genes involved in endocrine physiology could be transmitted to germline genes. This would lead to spontaneous and gene-specific alterations in gene function in the next generation. Another model to explain such phenomena involves altered somatic hormonal influences changing the 'regulatory settings' of germline genes. Hormone molecules in the circulation diffuse into the reproductive tissues, bind to hormone receptors on germ cells which in turn may directly alter the activity levels of the corresponding genes when they are expressed during the development of the embryo (advanced by Professor John Campbell of UCLA Medical School).[4]

INHERITANCE OF ACQUIRED CALLOUSING?

We now deal with some anatomical inheritance phenomena apparently wrought by experiments of nature. Some of these have been described in great detail by Professor Frederic Wood Jones in his 1943 book, *Habit and Heritage*. As an animal 'rubs itself up against the environment' as a consequence of repetitive habit, is there

190

evidence of an 'acquired inheritance' consistent with such
a habit? The examples we describe here involve apparently trivial phenomena yet they stretch our imagination
and make us wonder whether a strict neo-Darwinian
interpretation is sufficient.

Animals such as the ostrich and the African warthog
have large, prominent callosities in parts of the body
(sternum, forelimbs, hindlimbs) apparently as a consequence of their resting or foraging habits. The African
warthog forages by routing its large tusks into the ground
while kneeling on its forelimbs and pushing forward with
its hindlimbs. The strong, horny calluses appear to
protect the skin surfaces upon which it kneels. Ostriches
rest by squatting on their legs and their breast-bone
(sternum). Callosities can be found at all those sites
where the skin makes contact with the ground, the sternal
callus being particularly large. In these, and in many
other animals, including man, callousing can also be
induced to occur in other parts of the skin if subjected
to frequent rubbing. Callousing of this type can therefore
be classed as an 'acquired' somatic adaptation. What is
particularly interesting is that all those prominent natural
calluses found in adult ostriches and warthogs are *already
well formed in the embryo* in the absence of friction or
rubbing. This implies that these strategically located
callosities are *germline encoded*.[5] In the older literature,
and to zoologists such as Wood Jones, these observations
provide sufficient and compelling proof of the inheritance
of an acquired characteristic. To Wood Jones this is a
commonsense interpretation. The older neo-Darwinian
reaction (i.e. by the followers of Charles Darwin, not
Darwin himself) to these observations is that they are
peculiarities of a germline origin brought into being by
the natural selection process. A simple neo-Darwinian
explanation can be framed to account for the inherited
callosities. Because callousing can prevent skin damage

191

and thus reduce the risk of infection, the calluses could represent a clear selective advantage.

To 'Lamarckians' this has been construed as a glib explanation with inherent contradictions defying commonsense. This apparent 'flight from reason' by neo-Darwinians provoked the scathing published criticisms by nineteenth-century Lamarckian commentators such as Samuel Butler and the analyses published this century by Frederic Wood Jones and Arthur Koestler.

The only way to resolve this issue would be by controlled experimentation involving a breeding program (inducing callousing at an atypical skin site in each generation). Almost certainly this experiment is impractical for the very reasons articulated by Wood Jones. The other problem for a Lamarckian interpretation is that it is difficult to see how, based on present knowledge, such anatomically specific structural/morphological information could get back to the germline. It would be necessary first to document a soma-to-germline transmission route as well as to identify the genes involved. This makes the problem of the experiments with V genes look simple!

Thus callousing of skin is a visible and readily understandable phenomenon. Yet as we have seen, trying to resolve whether it is based on a 'Lamarckian' and/or 'Darwinian' evolutionary process is quite complex as it cannot be analysed easily by experiment. In contrast, antibody V genes cannot be seen yet they can be analysed and experiments can be done.

INHERITANCE OF ANATOMICAL PECULIARITIES OF THE SQUATTING HABIT IN HUMANS

In his book *Habit and Heritage*, Wood Jones also discussed variation in the squatting habit in different human populations. Many Asian peoples squat with their feet flat on the ground. However, Australian Aborigines have been recorded as squatting differently, with their feet tucked under their buttocks (Figure 7.1). In the Asian posture there are clear facets of bone structure (Figure 7.2) at the joint of the tibia (leg) and astragalus (foot) bones which seem to allow comfortable squatting in rice paddy fields and market places in that position for long periods (which most Caucasians find difficult to do). These particular squatting facets are not present in the Australian Aborigines, who have different facets as a result of their particular squatting habits. Further, both types of squatting facets are absent in those people who habitually use chairs or similar furniture on which to sit. What is most intriguing about all these squatting facet phenomena is that they are *already present* in the bones of embryos or young children before they develop the squatting habit. They are not present in the embryo or newborn of peoples who use chairs. The neo-Darwinist position must invoke a selective advantage (i.e. increased reproductive success) as a result of spontaneous germline mutations which affect the shape of the bones of the leg and feet. The key question is one of feasibility, given that these anatomical features must have arisen very recently in evolutionary terms.

These are interesting phenomena. The challenge in the future is for the design and execution of novel experiments which bear directly on the inheritance mechanisms.

ACQUIRED INHERITANCE IN BACTERIA?

Here we also want to briefly mention a controversial phenomenon first reported in *Nature* in 1988 by Professor John Cairns and co-workers of Harvard University. Without going into the details, they were able to generate gene mutations in cultures of bacteria, seemingly in a 'directed'

Figure 7.1 Squatting postures of Oriental races and Australian Aborigines

a.   b.

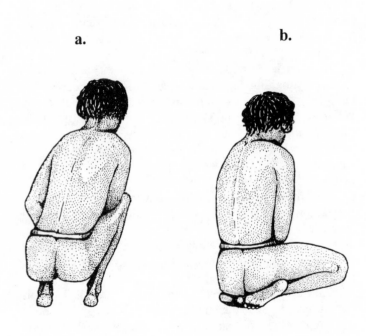

a. Oriental races; **b.** Australian Aborigines. (Reprinted from F. Wood Jones 1943 *Habit and Heritage,* with the permission of publisher, Kegan Paul International Ltd.)

Figure 7.2 Bone facets and squatting posture

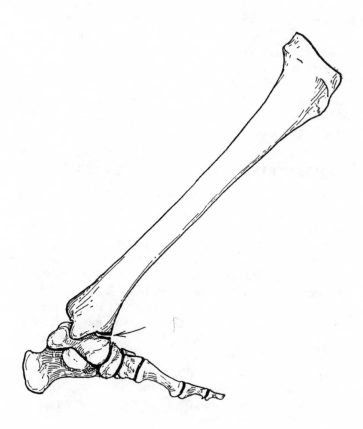

Bones of the left leg and foot to show the facets (indicated by the arrow) produced on tibia and astragalus by the Oriental posture of squatting. (Reprinted from F. Wood Jones 1943 *Habit and Heritage,* with the permission of publisher, Kegan Paul International Ltd.)

fashion. They showed that certain classes of mutation only appeared if chemical substrates related to the enzyme-encoding gene that mutated were present in the growth medium. These experiments have been controversial because other scientists have generated experimental results which have given rise to a conventional neo-Darwinian explanation, i.e. substrate selection for functional mutants with an efficiency advantage.[6] However, John Cairns' original explanation for how the mutants arose is intriguing; he invoked a reverse transcriptase-based gene feedback loop previously articulated by Ted Steele for the variable genes of the immune system.[7]

ACQUIRED INHERITANCE IN PLANTS?

What about acquired inheritance in plants? Here there is *no* 'Weismann Barrier' separating the soma and the germline. Acquired somatic modifications in plants which are gene-based can in principle be propagated to progeny when the seed is formed from that part of the plant that developed the somatic mutation.[8] So the secret is out— Lamarckian evolution is and has been, a fact of life in plants! Phenomena such as acquired inheritance of induced heavy metal tolerance can be routinely demonstrated.[9] The flax plant has been particularly useful in demonstating the role of environmental stress triggering inherited changes in plant genomes;[10] and the validity of this conclusion is underlined by the pioneering work by Nobel laureate Barbara McClintock on transposable genetic elements in maize.[11]

BEYOND THE IMMUNE SYSTEM—CAN RETROFECTION BE GENERALISED?

The picture of 'Lamarckian' gene feedback we have painted throughout this work has a clear and compelling logic for the origin and maintenance of the V genes of the immune system. Can we generalise the reverse transcriptase-based soma-to-germline idea for genes involved in other functions?

We should state first that some of our colleagues who now accept the weight of the evidence in the case of V genes would still want to confine the process *entirely* to the V genes. Thus our colleague Dr Arno Mullbacher (of the John Curtin School of Medical Research) has contributed significantly to the theoretical debate and experimental analysis of the V gene sequence data pertaining to soma-to-germline feedback by advancing critical and novel neo-Darwinian devil's advocacy scenarios. We are indebted to him, because his critical arguments have contributed materially to the success of the research program. Yet Arno stops short of accepting that the concept might be extended to genes responsible for other functions, or to other tissues and organs.

From the informal feedback we have received at seminars and conferences, Arno's position may be representative of many other scientists at this point in time. Most would baulk at speculating beyond the immune system. They may accept, albeit reluctantly, that the idea of 'Lamarckian inheritance' (retrofection of the germline by somatic genetic information) is now plausible for V genes, but not others.

Our rejoinder to these scientists is the 'thin-edge-of-the-wedge' argument. If one surveys the universe of biological phenomena one finds that apparently novel processes or functions are rarely unique. Once a precedent has been established for a structure or function in one living system, sooner or later someone will come across the same characteristic (or an analogue of it) in other cells, tissues or organisms. For example, not so long ago (early 1970s) reverse transcription was believed by many scientists to be restricted only to the C-type

RNA tumour viruses initially characterised by Howard Temin (in the face of sometimes virulent criticism). But we now know that RNA→DNA copying is widespread throughout viral and non-viral biological systems from bacteria to humans and the plant kingdom. A similar story can be recounted for the discovery of the widespread existence of the enzyme-like RNA catalysts (the so-called 'ribozymes') which were unknown prior to 1982 when they were first described in bacteria and protozoan unicellular organisms by Thomas Cech and Sydney Altman. As we will see, non-viral, *generalised* reverse transcription has had a significant and general impact in shaping the DNA sequence structure of higher vertebrate and mammalian genomes, quite apart from the role we have proposed for V gene evolution.

THE ROTHENFLUH MEMORY B LYMPHOCYTE MIGRATION MODEL: IMPLICATIONS FOR HOUSEKEEPING GENES

Can somatic cells penetrate the reproductive cell compartment and thus make close physical contact with germ cells? After reviewing the literature, Harry Rothenfluh has speculated that many non-random steps of the soma-to-germline feedback process could be effected by memory B lymphocytes, which would express mutated, antigen-selected rearranged variable region V(D)J genes. These cells could migrate into male or female reproductive tissue, enabling the delivery of somatically mutated variable region DNA sequence information to male and female gametes (sperm, ova).[12] These steps may or may not require endogenous retroviruses acting as gene vectors or sources of reverse transcriptase enzyme.

A possible consequence implicit in the Rothenfluh model is that migrant B lymphocytes might occasionally provide a general transport mechanism for the delivery of DNA sequences of *other* genes to the germline.[13] Evidence discussed in Chapter 2 suggests the possibility that endogenous retroviral reverse transcriptase could produce cDNA copies of cellular genes that still contain intron (non-coding) sequences. The cDNAs could be of any of those other non-antibody genes being expressed by the B cell and these potentially run into the hundreds, because they are both part of the families of 'tissue-specific' genes which define the functions of a lymphocyte and the 'housekeeping' genes which ensure the general viability of the cell. Most housekeeping genes are highly conserved with respect to the protein product they encode. Many of them are evolutionarily very ancient and are necessary for general cellular activity (e.g. genes encoding proteins involved in energy-generating pathways, the biochemical pathways that synthesise new membranes, the biochemical pathways that replicate the DNA of the chromosomes themselves, and the genes encoding histone proteins which are essential for packaging the long filaments of DNA into chromosomes).

By definition, therefore, unless a rare mutation benefits a housekeeping gene, such genes must display DNA sequences which *conserve the function* of the gene product (usually a protein). Retrotranscripts (cDNAs) would be expected to contain frequent sequence changes because of the error-prone copying from DNA to RNA and back again. A proportion of such base changes in the cDNAs would be so-called 'silent' or 'third position' base substitutions and would not influence the amino acid sequence of the encoded protein because of degeneracy in the triplet Genetic Code (see Appendix). The remaining base changes would change the amino acid sequence of the protein, usually with adverse effects on its function (these are called 'replacement' substitutions).[14] Such changes are most likely to reduce the fitness of any organism expressing the mutant gene and natural selection would eliminate them. However, genes with 'silent' changes would persist. Thus, the fact that housekeeping gene function must be conserved leads to a clear prediction: housekeeping genes are expected to display an excess of 'silent' base variations above the proportion expected from random unselected mutations. Such a pattern would be consistent with cDNAs copied from pre-mRNA being incorporated in the germline during evolution (although an elevated rate of 'silent' base changes could also be explained by a neo-Darwinian interpretation based on 'purifying' selection against germline mutations causing amino acid replacements).

Detailed analyses are now required of housekeeping genes in the light of the soma-to-germline idea. Several interesting predictions can be made and a very preliminary analysis suggests that these predictions will be fulfilled.[15] For example, a soma-to-germline model based on mobile cells predicts that the rate of base changes (mainly involving 'silent' substitutions) in housekeeping and tissue-specific genes expressed in B lymphocytes (and these would be many) should be higher than in those expressed in non-mobile cells. A rich source of data would be tissue-specific genes which code for the cytokines and cell surface proteins exclusive to lymphocytes. Within these genes, regions of the DNA which are transcribed into RNA and translated into protein should have higher evolutionary rates of base substitutions than the non-transcribed/non-translated regions. The model also predicts the strength of the 'Weiller' recombination signal[16] (presumptive soma-to-germline 'impact footprint') indicative of cDNAs being produced from pre-mRNA molecules (see Chapter 6) in various groups of homologous gene sequences. The 'Weiller signal' should be 'diluted' (less evident) for non-immunoglobulin gene families as these genes should not undergo somatic hypermutation. Therefore they should infrequently go through the error-prone DNA-to-RNA-to-DNA information loop.

There is now strong circumstantial evidence that soma-to-germline gene transfer involving reverse transcription has been going on over evolutionary time for *many* housekeeping genes. Because this evidence is indirect it could still have an alternative neo-Darwinian explanation. But it is the sort of molecular evidence *predicted and demanded* by a soma-to-germline theory involving the somatic feedback of cDNA retrotranscripts. Thus, a striking result from the isolation and DNA sequencing of many housekeeping genes over the past twenty years is that the majority exist as one (or at most two) functional copies per chromosome (which are transcribed into mRNA and then translated into a protein). In many cases, as a consequence of the isolation strategy, molecular biologists have carefully searched for closely related DNA sequences. These have been found, surprisingly, in relatively large numbers in the form of non-functional genes, termed 'pseudogenes', as they have been crippled by one or more disruptive mutations. But the really intriguing thing about these pseudogenes is that they have been copied from mRNA (after the non-coding intron sequences have been spliced out of the pre-mRNA) back into cDNA by the enzyme reverse transcriptase and are thus 'integrated retrotranscripts' or 'retropseudogenes'.[17]

The extraordinary picture we now have for many housekeeping genes in the mammalian genome is that the single-copy functional gene, which is highly conserved functionally, can be accompanied by many related retropseudogenes which exceed the functional gene in number by ten- to 100-fold. In many cases they are now located in other parts of the genome (e.g. on other chromosomes). To quote a recent authoritative text on molecular evolution:

201

Due to the ubiquity of reverse transcription the genomes of mammals are literally bombarded with copies of reverse transcribed sequences. The vast majority of these copies are non-functional from the moment they are integrated into the genome . . . [and] they are mostly located at great chromosomal distances from the parental functional gene . . . the phenomenon . . . has been likened to a volcano generating lava and the process has been termed the Vesuvian mode of evolution.[18]

This is a general feature for many higher animal species and in some cases thousands to hundreds of thousands of copies of retrosequences can be found littered all over the genome. The standard neo-Darwinian or neo-Weismann model (implied, never directly stated) is that some form of pre-mRNA synthesis using the DNA of the genes as templates is ongoing in germ cells themselves, producing multiple RNA 'copies' of a gene. Then intervention by a hypothetical reverse transcription process would produce many non-functional cDNA retrotranscripts which integrate at various sites in the chromosomal DNA of the germ cell in which they were formed. Such a model would require splicing of the pre-mRNA to remove introns, an event associated with normal gene expression (transcription and translation) if retropseudogenes are to be formed. This model requires that virtually all housekeeping and tissue-specific genes undergo transcription, splicing and reverse transcription in the cells of the germ tissues that produce sperm and eggs. We are not aware of evidence for this. It carries the risk that error-containing coding sequences could replace the standard gene by homologous recombination, in addition to the random 'Vesuvian' genome-wide integrations observed. Many of these new genes would be lethal or evolutionarily disadvantageous.

We detect such retropseudogenes because they are

damaged curiosities or 'molecular fossils'. Their wide-spread existence now underlines the possibility that successful soma-to-germline 'clean integration events' can also take place. A 'clean integration event' would arise via retrotranscripts being produced in somatic cells by reverse transcription from unspliced pre-mRNA templates encoding housekeeping genes. Such cDNA sequences may be mutated with 'silent' base changes and coding changes. Homologous recombination in the somatic cell to replace the original gene would allow functional testing of the new gene. Only somatic cells with successful 'silent' base changes (functionally conserved) would survive and multiply. Some of these mutant somatic cells could occasionally donate the new gene to germ cells either by migration (e.g. lymphocytes) or via a retroviral vector as discussed previously for antibody genes. However, the frequency of these events should be far lower than for antibody V genes, because only V genes undergo somatic hypermutation. The replacement of the original by the new gene by homologous recombination would go undetected as a soma-to-germline retrotransposition event because there is no way of distinguishing it from germline mutation and classical natural selection: the gene would be classed as an 'allele' (alternative functional variant) of the original gene.[19]

Therefore a general theory of soma-to-germline feedback states that cDNA sequences (both functional genes with introns and non-functional retropseudogenes) arise over evolutionary time in somatic cells and can be integrated into germline DNA in germ cells. The general theory predicts occasional 'clean soma→germline integration events' (homologous recombination) as ongoing throughout evolutionary history, maintaining the functionally conserved state of housekeeping genes (as the sequence has just been tested already for 'fitness' in a viable somatic cell), thus generating new functional alleles in the gene pool. However, non-functional cDNA sequences may be integrated in various sites in the genome, contributing to the growing number of retro-pseudogenes which accumulate over evolutionary time.

How do we choose between this 'neo-Lamarckian' model and the germline 'Vesuvian' model which is mandatory for a strict neo-Darwinian or neo-Weismann-ist paradigm? A more pertinent question would be: for what logical reason should we say that the proposed soma-to-germline mechanism is restricted only to the V genes of the immune system when the generalised nature of reverse transcription events shaping the DNA sequences in the genome for many genes is an established fact? Is there a logical reason therefore to prohibit the process? The evidence on the ubiquity and extent of genomic retrosequences on the face of it suggests there is no rational reason why the process should be quar-antined to only certain genes in the immune system.

In our opinion the balance of evidence discussed in this book supports soma-to-germline feedback rather than the 'Vesuvian' alternative. Thus it seems possible that soma-to-germline retrofection by cDNAs both maintains diverse open reading frames in germline V genes and also maintains the functionally conserved state of many housekeeping genes as well as tissue-specific genes.[20]

Retrotranscripts are not only a feature of animal and plant genomes. Viral genomes contain many examples of functional genes with high sequence similarity to cellular genes. The viral genes lack introns, a clear indication that they have arisen by reverse transcription from mature mRNA in host cells at some point in evolution. These genes are now employed in viral strategy during viral infection in animal hosts.[21] For example, they may encode an analogue of an animal protein that mimics the original and acts to blunt the effectiveness of the immune response, to the advantage of the virus.

The existence of such genes has led to speculation regarding trans-species (horizontal) transmission of genes by viruses which we discussed in detail earlier in this book (page 168).[22] To our knowledge, this speculation seems not to have been attacked by hardline neo-Darwinists and neo-Weismannists. This is extraordinary because it requires the penetration of Weismann's Barrier, i.e. a virus containing a gene from one species must infect the germ cells of another species in order to provide the possibility of integration of the new gene into the chromosomal DNA of the germline in the new host species. If this can happen between species, why not within the original host?

EPIGENETIC INHERITANCE: LAMARCKIAN DIMENSION

Our experimental and theoretical work on the possible 'retrogenetics' of the V genes of the immune system can be considered an example of a hard inheritance phenomenon involving genetic information embodied in the nucleotide bases of DNA sequences. There are, however, other forms of structural and functional inheritance in animals, plants and free-living unicellular organisms where the mechanistic steps do not involve direct alterations in DNA base sequence. There is a whole literature on these phenomena (e.g. the ubiquity of the phenomena in plants) which are outside the scope of this book. By saying this we are not downgrading their importance in contributing to our understanding of the origins of hereditary variations of evolutionary importance—our focus has simply been on the direct genetic penetration of the Weismann Barrier, as this provides the clearest example of a Lamarckian phenomenon which does not violate established genetic and natural selection principles. Recently this whole topic has been given an excellent scholarly treatment in a 1994 book by Eva Jablonka and Marion Lamb entitled *Epigenetic Inheritance and Evolution: The Lamarckian Dimension.*

FUTURE EXPERIMENTS

It is clear that rigorous controlled experimentation is needed to investigate the mechanisms of apparent 'acquired inheritance' phenomena. The new molecular technologies are providing us with unparalleled access to almost any gene(s) of interest and we have exploited DNA sequence information to search for the 'signature' of possible soma-to-germline feedback. However, this sort of evidence, irrespective of its intrinsic strength, will always remain indirect. The breeding programs needed to provide *direct evidence* require rigorous control and may take very large numbers of animals and a very long time. We are currently searching for clues as to the time-frame[23] by looking for changes in V genes of mouse lines originating from the same inbred (identical) genetic stock, but bred in different laboratories around the world for 40 years or more.

It is a matter now for 'Science' to put aside the dogmatic, ideological, emotional battles of the past and to focus on objective analysis. 'Lamarckian' ideas, as we have shown here, can play both a legitimate explanatory role as well as provoking experimental exploration of that class of inheritance phenomena associated with adaptive somatic responses such as immunity. We hope that the ideas and reasoning in this book provide a stimulus for further theorising and experimentation in fields foreign to us. The core of our approach is the fact that RNA can act as the template for DNA synthesis via reverse transcriptase enzymes, a fact so courageously established by Howard Temin in contradiction to the prevailing Central Dogma of molecular biology as it developed during the 1960s. Temin's work was predominantly concerned with retroviruses, but he recognised the broad implications for biology and evolution. We believe that his legacy may be more profound than even he could have imagined.

EPILOGUE

If we accept the feasibility of somatically acquired inheritance based on the molecular evidence for the immune system analysed so far, and if it came to pass that further scientific work revealed that the mechanisms discussed in this book could also be applied to other gene families, then there are some profound ramifications for science—and mankind. As the writing of this book was drawing to an end, it became apparent that some of these ramifications needed to be highlighted. Our purpose here, then, is speculative; we will examine the idea of a soma-to-germline feedback loop being a more general mechanism in the context of the preceding discussions, and to leave our readers with a few afterthoughts.

A RESPONSE TO NEO-DARWINIAN EXTREMISM

Charles Darwin and Alfred Russell Wallace first presented their joint theories of evolution based on the idea of natural selection to the Linnaean Society in London in 1858. Fierce opponents immediately branded the idea as destroying mankind's view of his relationship with God. As Daniel C. Dennett in *Darwin's Dangerous Idea* correctly points out, the storm surrounding this conceptual break with the past is still working its way through the intellectual, sociological, artistic and theological life of mankind. Prior to 1858 the intellectual challenge was to unite God with mankind and document 'His' role as the creator of the wider sphere of observable 'Nature'. The intellectual agenda was certainly never to suggest that man was merely a 'higher order animal', having descended from nematodes via the apes.

The Darwinian revolution was a resounding success, but the problem with intellectual revolutions is that they often harden into suffocating dogma—and at their apogee tend to be guarded by the holders almost as a sacred mantra. For a while the dogma is useful, but then an 'establishment' inevitably forms, the individuals of which find it almost impossible to break ranks, because if they do their careers and financial livelihood are put at great risk. This is antithetical to the spirit of scientific inquiry (yet sadly, an accurate commentary on the human condition). Today it is an unfortunate fact that *neo*-Darwinian ideas have evolved into almost a religion within sections of the scientific establishment. Neo-Lamarckian soma-to-germline gene feedback loops are still resisted with an irrational passion in a few quarters. This is particularly so amongst some immunologists and population geneticists who still seem wedded to the sterile concept, to us at least, of the 'neutral theory' of molecular evolution of Kimura.[1] Yet as we have shown here, the soma-to-germline concept has great explanatory and predictive power for the evolution and structure of the V-region gene families of the vertebrate immune system.

In 1998 the intellectual and scientific challenge is to deploy the scientific tools we now have at our disposal to determine whether the soma-to-germline concept can be extended beyond the immune system. We showed in Chapter 7 that all present indications are that it can be applied in a very fruitful way. This is a step forward, because we ought not forget that the battle for a certain legitimacy of some of Lamarck's ideas has been bitterly fought for more than 100 years. We gave a hint of some of this in the first chapter.[2]

One decisive denouement in 'Lamarckian' thinking took place in the first few decades of this century with the suicide of Paul Kammerer. The mid-century revival in Lamarckian thinking can be generally traced first through Frederic Wood Jones, then to H.G. Cannon and then Arthur Koestler through the 1960s and 1970s. Over the past twenty years the concept of soma-to-germline flow of genetic information involving mutated antigen-selected V-region retrotranscripts has had its own turbulent history (the full non-scientific story of which is too complex to recount here). The intensity of the opposition to this idea within both immunology and the wider sphere of biological science was so strong that by all 'the normal rules of the game' Ted Steele should not be working in science! For this reason, we are strongly motivated to address certain issues arising from the widely read writings of Daniel C. Dennett (a philosopher from Tufts University) and Richard Dawkins (of Oxford University), both of whom have emerged as late twentieth-century defenders of an extreme and uncompromising brand of neo-Darwinism. Stephen Jay Gould has recently called it 'Darwinian Fundamentalism',[3] and it has been also brilliantly exposed by the mathematician and novelist David Berlinski in his recent essay in *Commentary* magazine entitled 'The Deniable Darwin'.[4]

We first became aware of the extent of this extremism when reading Dennett's 1995 book *Darwin's Dangerous Idea*, which then prompted us to read an earlier book by Richard Dawkins, *The Extended Phenotype*. Dennett seems to rely heavily on Dawkins' biological and genetic judgement.[5]

In a chapter entitled 'Controversies Contained',

Dennett provocatively lists the Jesuit palaeontologist Teilhard de Chardin, Lamarck and 'directed mutations' as what he called 'the three losers'. Here we will focus solely on his short treatment of Lamarck. It is cursory and naive, although he acknowledges that Darwin himself 'notoriously' employed Lamarckian explanations based on his Pangenesis theory of use and disuse. Dennett completely accepts that Weismann's Barrier is genetically impermeable. He then argues against the possibility of acquired inheritance by stating '. . . for Lamarckism to work, the information about the acquired characteristic in question would somehow have to get from the revised body part, the soma, to the eggs or sperm (the germline). In general, such message-sending is deemed impossible— no communication channels have been discovered that could carry the traffic . . .'

This dogmatic statement is antithetical to everything we have discussed on how reverse transcription and somatic mutation may shape the V genes of the immune system. Indeed, as far as we can ascertain, reverse transcription and somatic hypermutation are not mentioned anywhere in his book. We grant this could be a knowledge void, as he is neither a molecular biologist nor immunologist. Nevertheless, he has positioned himself publicly as a knowledgeable and thoughtful 'evolutionary game player'. On closer scrutiny it became apparent that Dennett's assertions are similar to Dawkins' earlier assertions on the issue in his 1982 book *The Extended Phenotype*. We were intrigued by this because it was in the early 1980s that the earlier controversial episodes endured by Ted Steele took place following the publication of the Somatic Selection Hypothesis in 1979.

This prompted us to carefully read Dawkins' 1982 book and particularly the section drawn to our attention by Dennett in his footnote 5 on page 321, namely 'A Lamarckian Scare' (Dawkins 1982 pp. 164–178). As we briefly mentioned in Chapter 6, it is here that Richard Dawkins has dealt with the Somatic Selection Hypothesis at some length. Indeed Dawkins took it so seriously that he devoted the best part of ten pages to arguments attempting to deflate and defuse the central ideas of the theory. In the closing stages of his assessment on page 169, we were amused to read the following:

> Steele's theory, then, is a version of Darwinism. The cells which are selected, according to the Burnet theory, are vehicles for active replicators, namely the somatically mutated genes within them. They are active, but are they *germ-line* replicators? The essence of what I am saying is that the answer is an emphatic yes, if Steele's addition to the Burnet theory is true. They do not belong to what we have conventionally thought of as the germ-line, but it is a logical implication of the theory that we have simply been mistaken as to what the germ-line truly is. Any gene in a 'somatic' cell which is a candidate for proviral [read, 'retroviral'] conveyance into a germ cell is, by definition, a germ-line replicator. Steele's book might be retitled *The Extended Germ-line!* Far from being uncomfortable for neo-Weismannists, it turns out to be deeply congenial to us.

Dawkins' attempt to turn a lymphocyte and its genes into an 'extended germ-line' cell is scientifically absurd. In one breath he is accepting that a somatic cell *might* contribute new genes to the germ cells, a concept that is the antithesis of Weismann's dogma, and in the next breath he is saying that such a concept is 'deeply congenial' to neo-Weismannists. This has to be one of the all-time great back-flips only possible for a confirmed, dogmatic neo-Darwinist who has offered to eat his hat if neo-Lamarckism is true. If Weismann could read

212

that Dawkins paragraph, he would be spinning in his grave!

We have drawn attention to what Daniel C. Dennett and Richard Dawkins have published about Lamarck and the soma-to-germline idea because they have both made strong claims to large global audiences. And they have—for a time—got away with it. However, to reiterate our main point, the phenomenon whereby the defenders of orthodoxy adopt irrational positions has been very common in the history of science. The most famous case ushered in the modern scientific era during the Copernican revolution, when Galileo Galilei's academic and religious detractors refused to view for themselves the moons of Jupiter by looking at them through the telescope Galileo set up especially for the occasion. Then there was the ludicrous attempt by the catastrophist school of Cuvier of the late eighteenth and early nineteenth centuries to reconcile special creation and fixity of species with the geological evidence. Even at that time the fossil record (e.g. seashells on the tops of mountains) appeared consistent with catastrophic earthly upheavals. However, the accepted time since the creation was only about 6,000 years. This meant either that the previous 6,000 years were in constant upheaval or it had to be assumed (by Cuvier et al.) that 'each day was a period of indefinite (or variable) lengths'.[6] There are also fashion 'booms' in contemporary scientific disciplines that claim quantitative and molecular expertise. At times, large numbers of scientists are, for a while at least, held in thrall by erroneous concepts and experimental artefacts. Within the field of immunology there have been many such episodes, the most famous being the slow death of the instructionist theory of antibody formation years after Burnet's Clonal Selection Theory was published.

213

SPECIATION AND CONVERGENCE?

We would like to digress now to an area not yet mentioned by us, namely the implications of Lamarckian soma-to-germline feedback for our ideas on speciation and convergence. Apart from catastrophic extinctions possibly caused by meteorite impacts or widespread volcanic activity, the most perplexing phenomena in the evolution of life are the origins of new species and body plans—that is, the emergence of new and distinguishable life-forms. Through his eloquent books, particularly his *Wonderful Life*, Stephen Jay Gould has comprehensively shown how contingent the history of life really is. The 'Cambrian explosion' of multicellular life (approximately 570 million years ago) was the origin of many of the extant multicellular body plans we see today. Massive decimations of diverse life-forms occurred during this geological period. Indeed, the apparent fine-tuning of the surviving body forms and species since such 'adaptive radiation' events supports the theory of 'Punctuated Equilibrium' proposed by Gould and his colleague Niles Eldredge in 1972 (long periods of apparent stasis, or evolutionary constancy, being punctuated by explosive evolutionary bursts). An apparent feature of adaptive radiation events is the lack of many intermediate forms. Michael Denton has reviewed the reliability of this negative evidence at length in the mid-1980s in his book *Evolution: A Theory in Crisis*. Can a general theory of soma-to-germline transmission coupled to the concept of 'use and disuse' add anything to the debate on the origin of new life forms? Is it possible to conceive of complex somatic adaptations being translated into complex new morphological/physiological forms (or 'species'). That is, can many somatically mutated RNA/DNA sequences flow down the soma-to-germline 'channel' in a short time period (one or a few generations)? These events may be triggered by significant environmental change or upheaval. The necessity is to produce the potentially adaptive forms rapidly (obviously in one generation if the complex new line is to survive). This could be the mechanism of generation of a 'hopeful monster' discussed in the 1940s by the evolutionary geneticist Richard Goldschmidt in his book *The Material Basis of Evolution.*

However, soma-to-germline concepts can only contribute as *possible* adaptive diversifying mechanisms at the genesis of adaptive radiations. Other concepts such as 'horizontal gene transfer' between species will need to be invoked in the reconstruction of life histories of phyla and genera (by viral vectors or simply by 'gene ingestion' via the food). Indeed, Hoyle and Wickramasinghe in their book *Our Place in the Cosmos* provide a strong case for the large evolutionary step being caused by the arrival on Earth of new infectious genetic material from elsewhere in the solar system (see Figure 1.4). Of relevance also here are the very interesting ideas of the Santa Fe Institute's Stuart Kauffman on how complex ordered patterns can spontaneously emerge through self-organisation 'in enormous, randomly assembled interlinked networks of binary variables', resulting in what he calls 'order for free'.[7] And indeed the seminal theory by Lynn Margulis of endosymbiosis (swallowing without digesting) involving cell fusions[8] probably also played a decisive role in the diversification of the higher eukaryotic cells where each intracellular organelle probably enjoyed an independent cellular lifestyle but has now become a totally dependent symbiont for the greater good of the cell. It is conceivable that the phenomenon of apparent independent 'convergence' of form or physiology in quite unrelated species (indeed in different phyla) may be covered by the explanatory domain of a general theory of soma-to-germline information flow. But again such historical reconstruction would need to be coupled with

215

cooperative interactions (horizontal 'cross infections') be-tween genes and cells as just outlined. For an established species, an environmentally sensitive soma-to-germline feedback loop may have its greatest evolutionary value in allowing rapid and efficient 'genome tracking of the environment'.[9] In other words, during long periods of stasis Lamarckian soma-to-germline feedback may result in *adaptive fine-tuning*, allowing the species to become highly specialised in its niche.

GENETIC ENGINEERING

Another area where there are profound ramifications, if a possible neo-Lamarckian soma-to-germline feedback is found to be a more general mechanism, is in the brave new world of 'eugenic' or eukaryotic gene engineering. We have entered, for better or worse, a world in which DNA sequences are routinely 'copied, cut and pasted' from one cell to another.[10] The implications of this scientific revolution are now being explained in a number of widely available books (e.g. see Philip Kitcher's recent *The Lives to Come*). James Watson and Francis Crick started it all 45 years ago when they worked out the double helix structure of DNA. Somatic gene therapy is now being developed to cure 'inborn' genetic errors. We are right on the verge of widespread medical application of this technology. If a soma-to-germline 'genetic chan-nel' exists (albeit at this stage for V genes), is there any possibility that a 'somatic fix' of this type will also fix the germline?[11] Many of the corrected genes could be ferried into cells and integrated into the genome of the nucleus of the patient's white blood cells using specially modified retroviral gene vectors prior to transfusion back into the patient![12] The white blood cells targeted for receipt of these genes are the haemopoietic stem cells which turn over in bone marrow continuously replenishing the populations of all other types of blood cells—in humans millions of new blood cells per day are generated. This means that mobile white blood cells, which include the B lymphocytes, carrying and expressing RNA messages of the new functional gene, could penetrate reproductive tissues and potentially donate the new gene sequence to the germline DNA (as discussed in Chapters 6 and 7). Once again the unknown factor is the probability of such a transfer.

What we have discussed here is the strong possibility that somatic mutations within the immune system may be inherited in the germline. One main reason for thinking now in general terms is the 'thin-edge-of-the-wedge' argument, the view, outlined in Chapter 7, that once a process or mechanism appears in one biological system it is used in others. However, also in Chapter 7 we found it necessary to place limits on the possibility because the 'locus-specific device' controlling antibody variable gene mutation rates would probably be involved in ensuring high rate mutation only to that site in the genome. The logical extension would be that somatic hypermutation at other loci would need to possess such a regulator or 'Ei/MAR' analogue. Is it possible that high rates of somatic mutation have been overlooked in other 'adaptive' genetic loci such as the drug detoxifying systems of the liver (the cytochrome P–450 system)? Can we identify other possible 'sensory' gene systems as candidates, such as the olfactory genes? Indeed, under conditions of sustained somatic stress, in the target tissue is there stability in such 'sensory' genes (the gene products of which interact directly with those molecules providing the external environmental stimuli)? From this perspective, the emergence of tumours may well be the downside of somatic genetic adaptations—or the 'price of progress'—in multicellular eukaryotic organisms mounting an adaptive somatic response.[13]

217

Here, we can also introduce the concept of the 'selection window' in the context of the evolutionary origin of say cancers and autoimmune diseases. Interestingly enough these conditions are, by and large, diseases of post-reproductive age. Indeed, it is only in this century that some of the more affluent members of the human race are living long enough to develop them. Any somatic adaptive responses that may ameliorate them and be passed to the germline genes within an individual post-reproduction would therefore be lost to future generations. The general argument can be put that there has been little evolutionary selection against them because most of our ancestors began producing children from their healthy teenage years and died at what we would now call early middle age.

In affluent Western society there is a new phenomenon in human reproduction in that people are reproducing at an ever-increasing age. In turn this means there is a longer time window in which somatic genetic faults could be transmitted to germline genes. As a consequence the offspring of much older parents may be at a genetic disadvantage. And alternatively, some somatic adaptations arising from, say, 'lifestyle choice' may also confer a positive genetic advantage on the offspring of older parents. Given that this conclusion is contingent on the generality of the soma-to-germline concept, the probability of such events is unknown.

CONSCIOUS EVOLUTION

The emergence of consciousness is the most significant result of the evolution of life on earth. Even if the conscious brain has arisen by 'sheer dumb luck' it is now a fact of life and therefore must be factored into any scenario for the future evolutionary path of mankind. It is a tautology we cannot avoid. Its importance is

218

reflected by a new journal devoted entirely to the topic, *Journal of Consciousness Studies,* and the interesting ideas on 'human-centred mental realism' and the general role of consciousness in the evolution of the universe by Gary Richardson, a philosopher from the Blue Mountains, Australia. At the very least, man's conscious technological mastery over genetic tools has social and ecological implications.

What about somatic selection coupled to a soma-to-germline feedback loop? If it can be generalised, there are several consequences. The first is its role in our repertoire of natural 'avoidance responses'. In the analysis of epidemiological data, genetic abnormalities in families or populations should be viewed through the prism of the 'environment-before *versus* the environment-after mating' outlined earlier.[14] We should therefore eliminate those work practices and living environments which are likely to lead to future genetic degradation. Enlightened public policies are already being implemented in this regard (it makes sense under either a neo-Lamarckian or a neo-Darwinian paradigm).

The ethical implications are also profound. According to the neo-Darwinian paradigm the vagaries of the external environment have provided the tests of random germline mutations that have shaped our future genetic legacy. If the general concepts arising from the somatic selection idea are validated experimentally for more families of genes, organismal or internal 'agents of change' will also need to be considered as factors influencing our future genetic endowment. If we accept that 'Weismann's Barrier' has been rationally if not empirically shattered then serious consideration needs to be given to the proposition that our ideas and philosophy of life may, through choices of lifestyle, influence our future genetic make-up (because the soma-to-germline channel as a material process would be 'ethically blind'). It is conceivable that we will gain not

219

only an enhanced understanding of ourselves through the genetic 'looking glass', but we may very well discover the ability to harness and oversee our genetic destiny. This concept goes way beyond the current 'gene therapy' technologies emerging from biomedical research—we clearly have the ability *already* to create new genetic matter. It requires us to project our thinking forward to a time when it may be possible to envisage how conscious lifestyle choice might ultimately be translated into genetic out-comes or 'familial eugenics'.

The point here is not to dwell on these issues, but to highlight the scientific possibility of a relationship between our ideas about how life should be lived and our natural genetic make-up. This possibility poses many ethical and philosophical dilemmas. Ethical beliefs are, after all, a social construct based on many factors. But what if our genetic make-up is in the final analysis also a 'social construct' arising from our consciousness? The two may become closely entwined, and over evolutionary time may act to mutually reinforce each other.

For the aficionados we raise the issue of 'final causes', a concept that a traditional neo-Darwinist might incor-porate into any form of Lamarckian inheritance. The short answer is that 'divine purpose' or 'final causes' are outside the scope of scientific work. However, it is ironic that in his attempt to eliminate all traces of 'divine intervention' or 'deliberate design' from his extreme brand of neo-Darwinism, Richard Dawkins, in books such as *The Blind Watchmaker*, sets himself up as the supreme 'artificer of design'! This is lucidly explained in David Berlinski's essay 'The Deniable Darwin'. Thus all of his computer-based models demonstrating the 'power of natural selection' depend ultimately on Dawkins setting *all* the selection criteria and the sequential (algorithmic) rules for the desired result of his selection program. Thus 'divine intervention' by the programmer is intrinsic to all computer-based evolutionary selectionist programs, much like the way molecular biologists routinely use 'reverse genetics' to deduce the DNA sequence from amino acid sequences of the purified protein (via the rules of the Genetic Code). In our view the only useful concept would be one of 'anticipatory purpose' or 'genetic responsibility'—scientific knowledge together with ethical and moral attitudes and lifestyle choice *may* impact on our future genetic endowment.

We now stand on the threshold of what could be an exciting new era in genetic research. We have the scientific tools to expand our investigation of possible 'Lamarckian inheritance' for other types of genes if, as a society, we choose to do so. However, the 'politically correct' thought agendas of the neo-Darwinists of the 1990s are ideologically opposed to the idea of 'Lamarck-ian feedback'—just as the Church was opposed to the idea of evolution based on natural selection in the 1850s! To what extent will such opposition inhibit investigation of what are, in their opinion, radical possibilities? The social and multicultural policies developed by most West-ern democracies over the last 50 years have been designed to break down discrimination based on genetic differences (e.g. race, sex, colour, disability). Will these policies also be a barrier to accepting new ideas about inheritance? Ultimately what will the possible new genet-ics bring to the scientific understanding of ourselves, our social values, philosophy, health, religion and political codes?

We are on the threshold of an exciting new era of intellectual understanding about ourselves. As geneticists continue to work towards answering some of the ques-tions raised in this book, the important philosophical questions will be: Can we consciously direct our genetic future? Will our evolving social dogmas allow a better genetic future, or act to erode genetic fitness of humans as the dominant species on earth? Even if the questions arising from the somatic selection concept have only just begun to be answered, at least we can now start considering these other questions in a new light.

Appendix

The Genetic Code

First position	Second position				Third position
	U	C	A	G	
U	Phe	Ser	Tyr	Cys	U
U	Phe	Ser	Tyr	Cys	C
U	Leu	Ser	Stop	Stop	A
U	Leu	Ser	Stop	Trp	G
C	Leu	Pro	His	Arg	U
C	Leu	Pro	His	Arg	C
C	Leu	Pro	Gln	Arg	A
C	Leu	Pro	Gln	Arg	G
A	Ileu	Thr	Asn	Ser	U
A	Ileu	Thr	Asn	Ser	C
A	Ileu	Thr	Lys	Arg	A
A	Met or Start	Thr	Lys	Arg	G
G	Val	Ala	Asp	Gly	U
G	Val	Ala	Asp	Gly	C
G	Val	Ala	Glu	Gly	A
G	Val	Ala	Glu	Gly	G

Note: In DNA, **U** = **T**

The twenty standard amino acids grouped according to their basic chemical properties are:

- *Neutral or nonpolar groups*—**Gly** (glycine), **Ala** (alanine), **Val** (valine), **Leu** (leucine), **Ile** (isoleucine),

223

Translation of a messenger RNA base sequence into an amino acid sequence

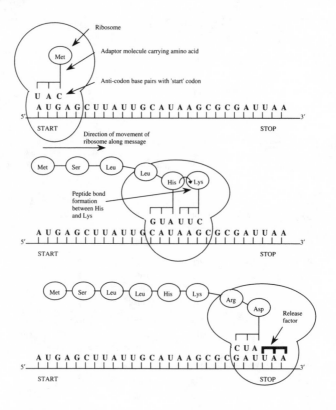

Met (methionine), **Pro** (proline), **Phe** (phenylalanine), **Trp** (tryptophan).

- *Neutral yet polar groups*—**Tyr** (tyrosine), **Ser** (serine), **Thr** (threonine), **Cys** (cysteine), **Asn** (asparagine), **Gln** (glutamine).
- *Basic or positively charged groups*—**Lys** (lysine), **Arg** (arginine), **His** (histidine).

224

- *Acidic or negatively charged group*—**Asp** (aspartate), **Glu** (glutamate).

We have grouped the amino acids this way because, as a rule of thumb, a mutation in a codon leading to the replacement of an amino acid can be 'conservative'. Thus a point mutation changing the codon **CUU** to **AUU** will replace leucine at that position for another neutral, non-polar amino acid, isoleucine, and such a change may partially conserve the function of the protein.

The synthesis of a protein is a very complex process involving tens if not hundreds of catalysed steps (both protein enzymes and RNA ribozymes appear to be involved). In the accompanying figure, a hypothetical and very short messenger RNA molecule (mRNA) is translated into an amino acid sequence. The 'molecular machine' carrying out the protein synthesising process is called a ribosome. Most proteins are 100–300 amino acids in length and are therefore specified by mRNA coding sequences of 300 to 900 nucleotide bases. The ribosome attaches and reads the message from left to right (in the 5' to 3' direction); as it moves down the message a specified sequence of amino acids (the 'protein') emerges from the ribosome into the surrounding medium.

Adaptor molecules (also called transfer RNA molecules or tRNAs) each with an amino acid attached are shown docking over the RNA message within the ribosome so as to align with the designated *codon* by the complementary base pairing of the adaptor's *anti-codon*. The amino acid sequence specified by the mRNA sequence is thus determined by the codon-to-amino-acid matching rules specified by the genetic code, i.e. the base sequence is read three at a time or in triplets. The ribosome therefore reads this linear message of triplet codons bringing together the specific adaptors with their particular amino acids so that a strong chemical bond forms between adjacent amino acids (the covalent or chemical bond is called a 'peptide bond'). In this way the amino acid chain grows in length.

When the ribosome comes to a *stop codon* it is prevented from extending the amino acid chain any further; a release factor binds to the stop codon causing the ribosome to release the newly formed protein into the surrounding medium. The mRNA and the ribosome can be used again to make another copy of the protein. The released amino acid sequence then folds into its particular three-dimensional shape dictated by its sequence of amino acids (but assisted by other proteins called 'chaperones').

The synthesis of a protein sequence can be very rapid: a chain of 100 amino acids can be synthesised under optimal conditions in about three seconds. In a long mRNA sequence several ribosomes can read the message at the same time (such a structure can be seen under the electron microscope and is called a 'poly-some').

GLOSSARY

ABO antigens The system of chemical molecules (antigens) on red cells which are recognised by naturally occurring antibodies (from another individual not of that blood type) and comprise the first level of classification of red blood cell (erythrocyte) antigens in the cross matching of 'donor' and 'recipient' blood for transfusion.

Affinity maturation Refers to the fact that mutant antibodies produced by the memory B cells which emerge from a Germinal Centre are of higher binding affinity for the antigen than the antibodies produced during the early phase of the immune response.

Alleles/polymorphism/isozyme Alternative DNA sequences of the same gene which change the structure of the protein slightly, yet conserve similar function are called *alleles*. A population of individuals expressing different alleles of the same gene is *polymorphic* for that gene. *Isozymes* are enzymes with the same basic function yet varied amino acid sequences which are expressed in *different* tissues.

Amino acids These are the basic chemical building blocks of proteins (see Table 1.1 and Appendix).

Antibody A protein produced by *B lymphocytes* (white blood cells) in response to foreign antigens (bacterial cells, virus particles and their toxin products). The antibody binds to the antigen and helps rid it from the body. A typical structure is represented by immunoglobulin G *(IgG)* a protein *heterodimer*—see Figure 3.2—consisting of two heavy *(H)* chains each with a molecular weight of 50,000 daltons (a protein of about 400 amino acids) and two light *(L)* chains each with a molecular weight of 25,000 daltons (about 200 amino acids in length). The basic subunit is a heavy + light, or *HL,* heterodimer—the antigen-combining site is composed of the variable regions of both H and L protein chains as indicated in Figure 3.2. Two *HL* units make up a complete *IgG* antibody; ten *HL* units make up a *pentameric IgM* molecule.

Antigen-binding site The intricate three-dimensional folding of the variable regions of an H and L chain constitute the surfaces of the heterodimer which bind the chemical moieties of the antigen. Also can be called *antigen-combining site.*

Antigen/'self antigen' An antigen is a large organic polymeric molecule, usually protein or a polymer of sugars (carbohydrate)—not a constituent of the 'self'—which stimulates the production of specific antibodies which bind specifically to the antigen and not other unrelated antigens. Typical foreign antigens are molecular constituents or products of bacteria and viruses. A self antigen is a protein or carbohydrate which is part of the body. Normal healthy animals and people do not make immune responses against self antigens.

Apoptosis The term used to describe a process of biologically controlled cell death. This is contrasted with other types of cell death (called 'necrosis') caused by outside causes (e.g. toxins, burns, trauma, etc.).

B lymphocyte Bone marrow-derived white blood cell which matures into antibody (Ig)-producing cells. B cells are mobile and circulate in the lymphatic fluids and blood. B cells can be bound and 'selected' by the antigen but they generally require an antigen-specific *helper T cell* for them to be activated into antibody secretion or cell proliferation. Each B cell expresses *one* antibody specificity.

Base Refers to the bases Adenine (**A**), Thymine (**T**), Guanine (**G**), Cytosine (**C**) in DNA and the base Uracil (**U**) which replaces **T** in RNA.

Base pairing Refers to the complementary base pairing of **A** with **T**, and **G** with **C** in DNA; and the pairing of **U** with **A**, and **G** with **C** in RNA (whether the latter is forming a duplex with a single strand of DNA or another single-stranded RNA molecule). See Figure 2.3.

Cap site The start site for mRNA synthesis.

Catalyst An agent, in biology an enzyme (protein) or ribozyme (RNA), which facilitates and speeds up a chemical reaction which would not normally take place if left to itself.

cDNA Complementary (copy) DNA sequence. This is the DNA sequence copied off an RNA sequence template by a 'reverse transcriptase' enzyme.

Chromosome A long double-stranded DNA molecule, circular in bacteria (and not complexed with proteins). It is linear in higher cells and complexed with 'histone' proteins, forming what is called 'chromatin'.

Clonal Selection As B cells mature from the bone marrow they express surface antibody (Ig) molecules all of the same specificity in any individual cell, i.e. the *same* sets of variable region genes are used in each Ig molecule. A foreign antigen entering the system selects, in a 'Darwinian' manner, only those B cells to which it can bind. These cells are activated

229

and divide (or proliferate), forming a *clone* of specific B cells, where all the progeny cells are *identical* in antibody specificity. This clone can then expand further in number with some of the progeny becoming *antibody-secreting B cells* and other progeny becoming long-lived *memory B cells* (these are the cells which are activated when the same antigen infects the individual at some later time—the principle of 'booster' vaccinations). Some memory B cells may also show evidence of *somatic hypermutation*—that is, the variable region (V) genes of the antibody have mutated away from the DNA sequence encoded by the germline V gene during the course of the immune response. The antibodies produced by such mutated memory B cells are usually of higher binding power (or affinity) than the antibodies made by B cells of the clone produced earlier in the immune response.

Coding region Refers to the sequence of DNA which actually codes for protein via the intermediary molecule, messenger (m)RNA.

Codon A triplet of three consecutive bases in mRNA which specifies a particular amino acid (see Appendix, Genetic Code).

Complementarity-determining region (CDR) The portion of the amino acid sequence in the variable region of an antibody combining site which makes physical contact with the antigen molecules. In comparing many antibody variable region sequences this is also the region which coincides with an elevated frequency of sequence variability. The regions are also called *HV, hypervariable regions*.

Constant (C) regions/immunoglobulin classes As the name implies, C-regions are those amino acid sequences of an antibody molecule which do not take part in antigen binding and are therefore not

230

subject to variation from one molecule to the next. The constant (C) regions of the heavy chain determine the function of the antibody, e.g. the ability to promote phagocytosis. C-regions of heavy chains can be classified into five general types of sequence (and therefore function) termed *immunoglobulin class* (below).

Cytokines/interferons   Protein molecules secreted usually by activated helper T cells which deliver activation and differentiation signals to other white blood cells including other T or B cells.

Cytoplasm   That part of the cell outside the nucleus containing the 'cell fluid' or cytosol and other organelles.

Diploid   Refers to two *haploid* sets of chromosomes in the same cell's nucleus (see Table 1.2). At fertilisation two gametes, one from the male (sperm) and the other from the female (ovum or egg), each carrying a 'haploid' set of chromosomes, fuse to form the fertilised zygote (which is now diploid).

DNA   Deoxyribonucleic acid (Table 1.1). The molecule which carries genetic information embodied in the sequence of deoxynucleotide bases, **A**, **T**, **C** and **G**. These are linked together by phosphodiester bonds. In its native form it is a very long base-paired polymer or a double helix consisting of millions of nucleotide bases (length depends on the species): it consists of two anti-parallel polynucleotide strands which are complementary base-paired, adenine (**A**) pairing with thymine (**T**) and cytosine (**C**) pairing with guanine (**G**). A single human chromosome consists of a very long double-stranded ('double helix') molecule of DNA consisting of tens of millions of bases.

DNA polymerase   Refers to the class of enzyme which makes DNA base sequence copies from a complementary template DNA base sequence. This set of enzymes is involved in the replication of the chromosome and in the repair of any base sequence errors that might arise during replication.

DNA synthesis or replication Refers to the synthesis of complementary copies of DNA synthesised in a 5′ to 3′ direction from another DNA template sequence.

Domain Used to denote a structural/functional unit of a protein, e.g. the constant or C-domain of immunoglobulins refers to the stretch of amino acids making up the constant region of the H or L chain.

Enzyme A protein which acts as a catalyst for a biochemical reaction.

Exon The amino acid coding sequences within a gene which are separated by non-coding interruptions called *introns*. These introns are 'spliced out' of the immature mRNA (pre-mRNA) to generate a mature mRNA molecule with all its *exon* sequences joined precisely together to form an in-frame contiguous message which is then read into an amino acid sequence by a ribosome (Figure 4.4).

FDC Follicular dendritic cell. Large spreading cells which form an extensive membranous network in a Germinal Centre. They present antigen–antibody complexes on their surface membranes and facilitate the affinity-based selection of B cell antibody mutants (see *Germinal Centre* and *Affinity Maturation*).

Five prime (5′) Refers to the 'left-hand end' of a DNA or RNA base sequence (Table 1.1).

Framework region (FW) These are conserved amino sequences (or nucleotide base sequences) within antibody variable regions. These regions contribute to basic structure of an antibody combining site but do not usually make physical contact with antigen molecules.

Gene conversion Similar if not identical to 'homologous recombination' except that it formally is meant to denote a one-way transfer of genetic information, from a donor sequence to a recipient sequence.

Genetic recombination The breakage and reunion of two different double-stranded DNA molecules.

Genome Refers to the complete set of genes in a cell or organism (or virus particle).

Germ cell A cell produced by the reproductive organs (sperm, egg), containing a single set (haploid set) of chromosomes.

Germinal Centre Special areas within a lymph node or spleen where antigen-activated B cells undergo intense proliferation (cell division), the cells dividing rapidly every 5–7 hours (in contrast to cell division times of up to 20 hours for most other somatic cells). A typical Germinal Centre would consist of up to 20,000 progeny B cells derived from one to three founder B cells. The process of somatic hypermutation is activated in Germinal Centre B cells. Those displaying *high affinity* mutant antibodies on their surface membrane are 'selected' by antigen binding to live and produce progeny *memory B cells;* B cells displaying non-functional or low affinity mutant antibodies are not selected by antigen binding and spontaneously die (by the process of *apoptosis*). The Germinal Centre therefore can be thought of as a site of intense Darwinian survival of the fittest antibody-producing B cells.

Germline Refers to the genes encoded in the chromosmal DNA of the germ cells.

Germline configuration Refers to the *germline variable* or *V gene element* which has not been rearranged to J or D/J elements. Since rearrangement only takes

233

place in lymphocytes, all other somatic cells and germ cells (sperm and eggs) have the germline configuration.

H chain Heavy chain of antibody molecule.

Haploid A single set of chromosomes, as opposed to *diploid*, where two homologous sets of chromosomes coexist in the same cell's nucleus, typical of a body cell in a diploid multicellular animal or plant. See Table 1.2.

Haptens Chemically synthesised antigens. Small chemical groupings, e.g. ring structures of carbon and nitrogen (benzene rings and their derivatives) prepared by chemists which by themselves are not able to induce antibodies. However, if haptens are physically bonded to larger 'carriers', such as foreign proteins, the hapten–protein complex can be used to immunise an animal and induce both anti-hapten and anti-protein antibodies.

Helper T cell A class of T lymphocytes which, when activated (usually by antigens presented in the context of MHC molecules) secretes cytokine hormones, which 'helps' the activation and growth of B lymphocytes and other T lymphocytes.

Heterodimer A complex protein consisting of two different protein subunits. Antibody and T cell receptor molecules are referred to as heterodimers because their basic protein structure is an H chain + L chain combination (antibodies) or a combination of one alpha (α) and one beta (β) chain (T cell receptor).

Histones In higher cells the very long DNA strands of the chromosomes are bound to histone proteins in ordered loops which allow packaging of the DNA into the small space of the nucleus.

Homologous gene A gene which across many species has the same or very similar function.

Homologous recombination The lining up of two homologous double-stranded DNA molecules resulting in genetic recombination—thought to be facilitated and guided by complementary base pairing.

Homologous sequence When referring to a DNA or protein sequence it means that the sequences have a high degree of similarity—probably greater than 70 per cent of the bases or amino acid residues are conserved at each position.

Housekeeping gene Refers to the large range of all *necessary* genes whose protein/RNA products are *essential* for the basic functions of the cell, for example, the enzymes of the energy-generating pathways, the enzymes needed to replicate the DNA, the enzymes and RNA molecules necessary to assemble the ribosomes and allow their function (see Appendix), the enzymes needed to splice out introns in pre-messenger RNA molecules, etc. Such housekeeping genes are to be distinguished from *tissue or cell-specific genes*, which are typically expressed by cells carrying out a particular specialised function. For example immunoglobulin or antibody genes are tissue-specific genes typically expressed by B lymphocytes; α and β globulin genes are expressed in that lineage of cells which become red blood cells expressing haemoglobin, the oxygen-carrying molecule.

Hybridoma A normal B cell can be 'immortalised' by a technique of cell fusion with a continuously dividing B tumour cell to create a so-called 'hybridoma'. Such a cell behaves like a tumour; it can be grown indefinitely in tissue culture (or an animal) and it secretes antibody of only one specificity (that of the original B cell).

Ig Abbreviation for immunoglobulin.

IgM, IgD, IgG, IgA, IgE These are the different classes of immunoglobulin distinguished by size (number of H + L subunits) and by sequence differences in their heavy chain constant regions which allow them to have different functions (e.g., activate complement enzymes, attach to cell membranes).

IgV Immunoglobulin variable gene element (refers to those V genes in the germline, or unrearranged configuration).

Immunoglobulin Formal term for an antibody protein abbreviated as *Ig*. There are five 'classes' of immunoglobulin termed *IgM*, *IgG*, *IgD*, *IgA* and *IgE*. They are classified by differences in amino acid sequence in the C-regions of their H chains. The functions of the classes also differ. All have a membrane-bound form whereby they are displayed on the surface membrane of a B lymphocyte. *IgD* is involved in the antigen-triggering mechanism and is never secreted into the surrounding body fluid. *IgM*, *IgG*, *IgA* and *IgE* are all secreted from a B cell; the secreted form of *IgM* is pentameric (ten HL heterodimer units), appears early in an immune response and is a powerful promoter of phagocytosis; *IgG* consists of two HL heterodimer units and is secreted later in an immune response and is an antibody class typical of 'memory' responses. (These antibodies are usually of higher affinity, important in the binding and neutralisation of viruses and microbial toxins.) *IgG* antibodies produced by the mother can cross the placenta and enter the circulation of the foetus. *IgA* consists of two or four HL heterodimer units; it is the antibody type which protects the mucosal surfaces of the respiratory and gastrointestinal tract. It is also the main antibody in the early milk (colostrum). *IgE* consists of two HL

236

heterodimer units and is the antibody class responsible for acute allergies and hay fever.

Intron The non-coding sequence which interrupts the coding region of a gene. This is 'spliced out' of the immature mRNA (pre-mRNA) to generate a mature mRNA molecule.

Junk DNA/'flanking' region sequences Often a term used to refer to those parts of the chromosomal DNA sequence which do not appear to code for proteins or other types of functional RNA (ribosomal RNA, transfer RNA). Another term to describe the DNA sequences that surround coding regions is *'flanking'* sequences.

Killer T cell A T lymphocyte which recognises, via its T cell receptor (TCR), a virus-infected cell within the body, leading to the destruction of the target cell. Also called a 'cytotoxic' T cell (or *Tc*).

L chain Light chain of antibody molecule.

Locus-specific device A term used to describe a docking (binding) site for the RT-mutatorsome which allows the focusing of somatic hypermutation to *only* the V(D)J and its contiguous immediate DNA sequence.

MHC Major Histocompatibility Complex molecule. Binds and presents 'foreign peptides' for recognition by T cell receptors (TCRs).

mRNA Messenger RNA which is transcribed (copied) from the DNA base sequence of a gene encoding the amino acid sequence of a protein. The ribosome reads this message sequence three bases at a time (= codons) and specifies the sequence of amino acids in the protein (see Appendix); mRNA molecules (or 'transcripts') are therefore much shorter in length than the DNA molecule of the chromosome—an mRNA transcript is a copy of a specified region which can be translated into protein.

Mutagen   Any agent (usually a hazardous chemical or penetrating electromagnetic radiation) which can cause mutations (changes in a DNA base sequence) such that when the DNA is replicated the mutation in the base sequence is copied into progeny DNA molecules.

Nucleic acids   The chemical term for DNA or RNA.

Nucleotides   These are the basic chemical building blocks, which, with their attached 'bases' (**A**, **T**, **C**, **G** or **U**) when linked together by phosphodiester bonds into a polymer make up strands of DNA or RNA.

Nucleus   The membrane-bounded organelle within a higher cell which contains the complete set of chromosomes; for a somatic cell this would usually be a diploid number, and in a male or female gamete, this would be a haploid set.

Organelle   Refers to a complex molecular structure or 'organ' within the cytoplasm of a higher cell which has a specialised function within the cell. Thus the 'mitochrondrion' is the energy powerhouse of the cell (it produces 'ATP' molecules which carry high-energy phosphate bonds, the energy currency of the cell). In a plant cell the 'chloroplast' is the site where radiant energy from the sun is converted into chemical energy (sugar molecules) with the release of oxygen (so-called 'photosynthesis').

PCR   Polymerase chain reaction. A technique for the specific amplification of a region of DNA.

Peptide   A 'small' protein consisting of 10–20 amino acids linked together by peptide bonds. When a protein is degraded by digestive enzymes ('proteases') it is usually said to be 'broken down to peptides'.

Phagocyte/phagocytosis   Phagocytes are white blood cells which engulf (phagocytose) and digest bacteria and other cellular debris. To be phagocytosed, bacteria and viral particles may need to be coated with antibody molecules.

Point mutation Refers to a single base change. For example a nucleotide base **G** changing to any of the other three bases **T**, **A** or **C** (see Figure 2.5). The mutations observed in mutant antibody variable *(V(D)J)* regions are usually all of the point mutation type (as opposed to removal or addition of one or more bases, formally termed '*insertions or deletions*').

Protecton A unit of antibody-mediated 'protection' defined (by Langman and Cohn) as the minimum critical concentration of specific antibody in the blood required to protect against infection. It provides a rationale by which the minimum number of immune lymphocytes per millilitre of body fluid required to produce a protective concentration of antibodies in a specified time can be estimated.

Proteins A long polymer of amino acids linked by peptide bonds (Table 1.1), usually 100–300 amino acid residues in length.

Rearranged variable region gene Also referred to as *V(D)J*, which can apply to a rearranged H (VDJ) or L (VJ) chain variable gene. In this '*somatic configuration*' two basic events ensue: first, the V(D)J can be transcribed into mRNA and thus eventually translated into a protein chain (H or L); second, it can serve as the target for the *somatic hypermutation* mechanism.

Replicase/polymerase A generic term for a polymerase enzyme that copies DNA or RNA base sequences into progeny DNA or RNA base sequences. DNA polymerase (DNA replication) and RNA polymerase (transcription) and reverse transcriptase (copying an RNA template sequence into DNA) can all be classed as replicases.

Reverse transcriptase An enzyme which copies an RNA base sequence into a DNA base sequence.

Reverse transcription The synthesis of a complementary copy of DNA (termed cDNA) from a single-stranded RNA template by a reverse transcriptase enzyme. The synthesis is always in the 5' to 3' direction. Reverse transcriptase can then make the cDNA a double-stranded molecule.

Ribosome A molecular organelle made of RNA molecules and folded proteins which allows translation of an mRNA sequence into proteins. See Appendix.

Ribozyme An RNA molecule which can act as a specific catalyst, either to act on itself or another RNA, DNA, or protein molecule.

RNA Ribonucleic acid (Table 1.1). These are usually single-stranded complementary copies of selected stretches of DNA sequence. It is also made up of ribonucleotide bases, adenine (**A**), uracil (**U**), cytosine (**C**) and guanine (**G**). In RNA, **U** is functionally equivalent to **T** in DNA.

RNA intermediate In the context of this book and the concept of reverse transcription, RNA intermediate refers to the fact that a copy of a gene (DNA sequence) made into RNA (by transcription) can be reverse-transcribed into cDNA or a retrotranscript. In this sense, then, particularly if the cDNA becomes integrated into the chromosomal DNA, the gene is said to have been copied via an RNA intermediate.

RNA polymerase Refers to the class of enzyme which makes a complementary RNA base sequence from a DNA template sequence. These enzymes do not possess error-correcting or editing functions, and RNA synthesis is therefore very error-prone compared to DNA synthesis.

RT-mutatorsome The hypothesised molecular organelle within the nucleus of a B cell responsible for somatic hypermutation (Figure 5.6). RT refers to the enzyme reverse transcriptase which copies, in an error-prone fashion, the pre-mRNA of the V(D)J into DNA (termed a retrotranscript or cDNA). The mutated cDNA copy of the V(D)J is then integrated into the chromosome so as to replace the original V(D)J DNA sequence.

Soma-to-germline feedback loop Refers to a process whereby *somatically mutated V(D)J* genes are reverse-transcibed to *cDNA*, which then recombines with the most similar (homologous) germline V gene sequence so as to replace it (termed *homologous recombination*), see Figures 1.2 and 6.3.

Somatic cell A body cell of a multicellular organism (a cell which does not produce the sex or reproductive cells, the eggs or sperm).

Somatic configuration Refers to the *rearranged variable* or *V gene* which is *only* found in mature B and T lymphocytes. Such a variable gene is given the generic symbol *V(D)J*.

Somatic hypermutation Refers to the mutation process activated by antigenic stimulation of a B cell which causes the *rearranged variable region genes—(V(D)Js)*—of the antibody expressed in that B cell to mutate the bases in their DNA sequence at a rate a million times faster than the normal background mutation rate. The process is tightly controlled and occurs almost exclusively in special cellular micro-environments called *Germinal Centres*.

T lymphocyte Thymus-derived white blood cell, so called because the cell acquires its special functions by maturing within the micro-environment of the thymus gland (precursor T cells are originally from the bone marrow). T cells are mobile, circulating in the lymphatic fluids and the blood. They express antibody-like surface molecules called *T cell receptors*

241

(or *TCRs*) which have antigen-recognising V-regions similar in principle to the antigen-combining sites of antibody molecules. TCRs recognise complexes of foreign protein (broken into a small stretch of amino acids called a 'peptide') and the cell surface MHC molecule.

T cells come in two basic types: (a) T cells that kill virus-infected target cells and are termed *Tc*, cytotoxic or *killer T cells;* (b) T cells that 'help' the activation or function of other T or B cells and are termed *helper T cells (Th)*. Helper T cells mediate their function by secreting protein *cytokines* that provide 'activation' or 'differentiation' signals for other cells of the immune system.

Template molecule Any single-stranded DNA or RNA sequence which serves as the template for copying by a 'replicase'.

Three prime (3′) Refers to the 'right-hand end' of a DNA or RNA base sequence (Table 1.1).

Transcription The synthesis of a single-stranded complementary copy of RNA from a DNA template by an RNA polymerase enzyme. The chain is always synthesised in 5′ to 3′ direction.

Translation The synthesis of a specific sequence of amino acids in a protein by a ribosome directed by the sequence of specific codons in the mRNA sequence. See Appendix.

Transplantation antigens/killer T cells Also called *major histocompatibility antigens* or *MHC antigens*. Expressed on most cells of the body. Rare to find any human individual with the same combination of MHC antigens (except identical twins) and this explains why graft rejection is the rule rather than the exception. Antibodies to the ABO antigens on red blood cells help distinguish the major types of blood groups, and in a similar vein, *killer T cells* from one individual can recognise the MHC antigens of a non-identical individual and react to them, thus rejecting transplanted tissue. The MHC antigens also act as *antigen presenting structures*, e.g. MHC molecules of a virus-infected cell bind fragments of the viral proteins (peptides) and present the molecular complex of MHC + viral peptide to killer T cells, activating them and causing the killer T cells to then attack and lyse the virus-infected cell.

V(D)J Generic term for a rearranged heavy or light chain variable region of an antibody.

V gene Variable gene. Usually refers to the unrearranged V genetic element in the *germline configuration*.

Variable (V-) region/antigen-combining site Both H and L chains each contribute 'variable' amino acid sequences at their tips (see Figure 3.2). Also termed an *antigen-binding site*. V-regions form the antigen-combining site of an antibody (or T cell receptor). Each different antibody has a different antigen-binding specificity which is reflected in the different sequences of amino acids in their V-regions.

NOTES

CHAPTER 1 THE TWIN LEGACIES OF LAMARCK AND DARWIN

1 Alfred Russell Wallace independently arrived at the same general conclusion and published it at the same time as Charles Darwin. Wallace, much younger than Darwin, remained his lifelong friend and graciously gave Darwin scientific priority mainly because independent testimony confirmed that Darwin had spent the previous 20 to 30 years marshalling the evidence to support the theory prior to his simultaneous publication with Wallace.

2 J.B. Lamarck. *Zoological Philosophy* (1809). Translated by Hugh Elliot. University of Chicago Press, Chicago, 1984.

3 C. Darwin. *The Variation of Animals and Plants Under Domestication*. John Murray, London, 1868, see the penultimate chapter of the 2nd volume.

4 This and other earlier acquired inheritance data and experiments discussed in J.A. Detlefsen 'The Inheritance of Acquired Characters' *Physiological Reviews* vol 5: 244–278, 1925.

5 Z.A. Medvedev. *The Rise and Fall of T.D. Lysenko*. Translated by I.M. Lerner. Columbia University Press, New York, 1969.

CHAPTER 2 IN THE BEGINNING THERE WAS RNA

1 Lipids are molecules important to the structure and functioning of membranes. In this book we will not discuss much about lipids and biological membranes, or carbohydrates. These large polymeric and multi-domain molecules are no less important than nucleic acids and proteins, as they are integral parts of all cells including virus particles, but this is beyond the scope of the book.

2 This result confirmed the pioneering work (at the Rockefeller University) published in 1944 by Oswald Avery, Colin MacLeod and Maclyn McCarty. They showed that the factor responsible for the phenomenon of 'genetic transformation' in bacteria was associated with chemically pure DNA.

3 There is, however, a term often used by molecular biologists called 'reverse genetics'. Here the molecular biologist can deduce the DNA sequence from a sequence of amino acids, but this 'reverse translation' is done by reference to the genetic code (shown in the Appendix). We will not make a categorical denial of 'reverse translation' because it is always possible that an organism (in an unusual habitat?) will be discovered that can carry out the process.

4 F.H.C. Crick and L.E. Orgel 'Directed Panspermia' *Icarus* vol 19: 341–346, 1973.

5 P. Klenerman, H. Hengartner and R.M. Zinkernagel 'A Non-retroviral RNA Virus Persists in DNA Form' *Nature* vol 390: 298–301, 1997.

6 See T.R. Cech 'RNA as an Enzyme' *Scientific American* vol 255: 76–84, 1986. Reviewed also in R.F. Gesteland and J.F. Atkins (Eds). *The RNA World*. Cold Spring Harbor Laboratory Press, New York, 1993.

7 M. Eigen, W. Gardiner, P. Schuster and R. Winkler-

Oswatitsch 'The Origin of Genetic Information' *Scientific American* vol 244: 78–94, 1981.

8 In strict material terms this translates to 'complex cascades of interactions between nucleic acids and proteins'.

CHAPTER 3   WHY THE IMMUNE SYSTEM IS SO INTERESTING

1 D.R. Hopkins. *Princes and Peasants: Smallpox in History.* University of Chicago Press, Chicago and London, 1983.

2 F. Fenner, D.A. Henderson, I. Arita, Z. Jezek and I.D. Ladnyi. *Smallpox and its eradication.* World Health Organisation, 1988.

3 L. Pasteur 'Vaccination in Relation to Chicken-cholera and Splenic Fever' *Transactions of the International Medical Congress*, 7th session, London, 2–9 Aug 1881 vol 1: 85–90.

4 B.D. Davis, R. Dulbecco, H.N. Eisen, H.S. Ginsberg and W.B. Wood Jr. *Microbiology.* Harper & Row, New York, p. 505, 1969. Reprinted with the permission of the publisher, Harper & Row/Paul Hoeber (Medical Textbooks) Lippincott-Raven, Philadelphia.

5 G.W. Litman 'Sharks and the Origin of Vertebrate Immunity' *Scientific American* vol 275: 47–51, 1996.

6 M. Cohn 'The Molecular Biology of Expectation' In: *Nucleic Acids in Immunology*, O.J. Plescia and W. Braun (Eds) Springer-Verlag, New York, pp. 671–715, 1968.

CHAPTER 4   THE IDEA OF 'CLONAL SELECTION'

1 G.L. Ada and G. Nossal 'The Clonal Selection Theory' *Scientific American* vol 255: 50–57, 1987.

2 The point of the metaphor is to show that as the density of pixels or different antibodies increases, both the picture definition becomes clearer and the self vs non-self definition becomes easier to establish.

3 S. Tonegawa 'Somatic Generation of Antibody Diversity' *Nature* vol 302: 575–581, 1983.

4 R.E. Langman and M. Cohn 'The E-T (Elephant-Tadpole) Paradox Necessitates the Concept of a Unit of B-Cell Function: The Protecton' *Molecular Immunology* vol 24: 675–697, 1987.

5 C. Chen, Z. Nagy, E.L. Prak and M. Weigert 'Immunoglobulin Heavy Chain Gene Replacement: A Mechanism of Receptor Editing' *Immunity* vol 3: 747–755, 1995. We have good reason for thinking that this genetic recombination mechanism could be part of the soma-to-germline homologous recombination feedback loop, at least for H chain V genes.

6 For an overview see G.W. Litman 'Sharks and the Origin of Vertebrate Immunity' *Scientific American* vol 275: 47–51, 1996; and M.K. Anderson, M.J. Shamblott, R.T. Litman and G.W. Litman 'Generation of Immunoglobulin Light Chain Gene Diversity in *Raja erinacea* Is Not Associated with Somatic Rearrangement, an Exception to a Central Paradigm of B Cell Immunity' *Journal of Experimental Medicine* vol 182: 109–119, 1995.

7 See H.S. Rothenfluh, R.V. Blanden and E.J. Steele 'Evolution of V Genes: DNA Sequence Structure of Functional Germ-line Genes and Pseudogenes' *Immunogenetics* vol 42: 159–171, 1995. See also text and legend of Figure 6.2. We also acknowledge and thank Rod Langman for first pointing out to us that the 'fused D-bits' are in the preferred reading frame (there are three possible frames) and thus implying that they could have arisen only after a fully functional protein antibody could be tested by binding

an antigen. This type of evidence is discussed in Chapter 6 as strongly supportive of soma-to-germline feedback via cDNA retrotranscripts.

8 J.O. Kephart, G.B. Sorkin, D.M. Chess and S.R. White 'Fighting Computer Viruses' *Scientific American* vol 277: 56–61, 1997.

9 J. Mehra *The Beat of a Different Drum: The Life and Science of Richard Feynman* Clarendon Press, Oxford, p. 441, 1994. The concept is also explored in G. Milburn's *The Feynman Processor: An Introduction to Quantum Computation* Allen & Unwin, Sydney, 1998.

CHAPTER 5 SOMATIC MUTATION

1 We also include all those species of RNA molecule which are 'gene products', such as transfer RNA and ribosomal RNA (see Appendix).

2 For a review of the original literature on which this chapter is based see E.J. Steele, H.S. Rothenfluh, G.L. Ada and R.V. Blanden 'Affinity Maturation of Lymphocyte Receptors and Positive Selection of T Cells in the Thymus' *Immunological Reviews* vol 135: 5–49, 1993. For an update see E.J. Steele, H.S. Rothenfluh and R.V. Blanden 'Mechanism of Antigen-Driven Somatic Hypermutation of Rearranged Immunoglobulin V(D)J Genes in the Mouse' *Immunology and Cell Biology* vol 75: 82–95, 1997.

3 Summarised and reviewed in J. Foote and H.N. Eisen 'Kinetic and Affinity Limits on Antibodies Produced During Immune Responses' *Proceedings National Academy of Science USA* vol 92: 1254–1256, 1995. See also H-P. Roost, M.F. Bachmann, A. Haag, U. Kalinke, V. Pliska, H. Hengartner and R.M. Zinkernagel 'Early High-Affinity Neutralizing Anti-viral IgG Responses Without Further Improve-

ments of Affinity' *Proceedings National Academy Science USA* vol 92: 1257–1261, 1995; M.F. Bachmann, U. Kalinke, A. Althage, G. Freer, C. Burkhart, H-P. Roost, M. Aguet, H. Hengartner and R.M. Zinkernagel 'The Role of Antibody Concentration and Avidity in Antiviral Protection' *Science* vol 276: 2024–2027, 1997.

4 See the overview by P. Parham 'A Boost to Immunity from Nurse Sharks' *Current Biology* vol 5: 696–699, 1995; A.S. Greenberg, D. Avila, M. Hughes, A. Hughes, E.C. McKinney and M.F. Flajnik 'A New Antigen Receptor Gene Family that Undergoes Rearrangement and Extensive Somatic Diversification in Sharks' *Nature* vol 374: 168–173, 1995; and K.R. Hinds-Frey, H. Nishikata, R.T. Litman and G.W. Litman 'Somatic Variation Precedes Extensive Diversification of Germline Sequences and Combinatorial Joining in the Evolution of Immunoglobulin Heavy Chain Diversity' *Journal of Experimental Medicine* vol 178: 825–834, 1993.

5 DNA rearrangement to assemble a diverse somatic repertoire of H or L chain V(D)J variable region genes does not occur at all in avian species (chickens). These animals use a gene conversion strategy to diversify a single functionally rearranged H or L chain gene in a mature lymphocyte. Recently we have unified this gene conversion strategy with the RT-mutatorsome concept outlined in this chapter. We have shown that reverse transcription and homologous recombination could underpin both somatic gene conversion and somatic hypermutation (typical of chicken, but we also apply it to rabbit and sheep Ig loci), see R.V. Blanden and E.J. Steele 'A Unifying Hypothesis for the Molecular Mechanism of Somatic Mutation and Gene Conversion in Rearranged

Immunoglobulin Variable Genes' *Immunology & Cell Biology*, vol 76: 288–293, 1998.

6 See the review by M.S. Neuberger and C. Milstein 'Somatic Hypermutation' *Current Opinion in Immunology* vol 7: 248–254, 1995. Also see our own review of this evidence—E.J. Steele, H.S. Rothenfluh and R.V. Blanden 'Mechanism of Antigen-Driven Somatic Hypermutation of Rearranged Immunoglobulin V(D)J Genes in the Mouse' *Immunology and Cell Biology* vol 75: 82–95, 1997. For reviews of the somatic hypermutation field prior to the advent of genetically defined transgenic mice (prior to 1990), including some early history, see E.J. Steele (Ed.) *Somatic Hypermutation in V-Regions* CRC Press, Boca Raton, FL, 1991.

7 D.C. Reanney 'Genetic Error and Genome Design' *Trends in Genetics* vol 2: 41–46, 1986.

8 For the general reader see M. Radman and R. Wagner 'The High Fidelity of DNA Replication' *Scientific American* vol 259: 24–30, 1988.

9 Also called 'polymorphisms'. Often such variant gene sequences co-exist within normal healthy individuals, in which case, if they are enzyme-coding genes, they are called 'isozymes'. Isozymes carry out tissue-specific functions.

10 Reviewed in A.J. Cunningham 'Evolution in Microcosm: The Rapid Somatic Diversification of Lymphocytes' *Cold Spring Harbor Symposium on Quantitative Biology* vol 41: 761–770, 1977. An appreciation of Cunningham's work can be found in E.J. Steele 'Alistair Cunningham and the Generation of Antibody Diversity *After* Antigen' *Immunology and Cell Biology* vol 70: 111–117, 1992.

11 Reviewed in M. Cohn 'A Rationale for Ordering the Data on Antibody Diversity' In: *Progress in Immunology II: Biological Aspects*, I.L. Brent and

J. Holborow (Eds). North-Holland Pub. Co.,
Amsterdam, pp 261–284, 1974.
12 See E.J. Steele and J.W. Pollard 'Hypothesis: Somatic
Hypermutation by Gene Conversion via the Error
Prone DNA→RNA→DNA Information Loop' *Molecular Immunology* vol 24: 667–673, 1987. The evidence
supporting the model was subsequently reviewed to
1990 in E.J. Steele (Ed.) *Somatic Hypermutation in
V-Regions* CRC Press, Boca Raton, FL, 1991; and to
1996 in E.J. Steele, H.S. Rothenfluh and R.V. Blanden
'Mechanism of Antigen-Driven Somatic Hypermutation
of Rearranged Immunoglobulin V(D)J Genes in the
Mouse' *Immunology and Cell Biology* vol 75: 82–95,
1997.
13 R.V. Blanden and E.J. Steele 'A Unifying Hypothesis
for the Molecular Mechanism of Somatic Mutation
and Gene Conversion in Rearranged Immunoglobulin
Variable Genes' *Immunology and Cell Biology*, vol 76:
288–293, 1998.

CHAPTER 6 SOMA-TO-GERMLINE
FEEDBACK

1 A summary of the essential points of this chapter
can be found in two semi-popular articles written for
the general reader. See H. Rothenfluh and T. Steele
'Lamarck, Darwin and the Immune System' *Today's
Life Science* vol 5(7): 8–15 and vol 5(8): 16–22, 1993;
and E. Steele, R. Blanden and H. Rothenfluh 'How
Have Antibody Genes Evolved?' *Australasian Science*
vol 17(4): 46–49, 1996. Most of the detailed molecular evidence can be found in H.S. Rothenfluh, R.V.
Blanden and E.J. Steele 'Evolution of V Genes: DNA
Sequence Structure of Functional Germ-line Genes
and Pseudogenes' *Immunogenetics* vol 42: 159–171,
1995. See also the mid-1980s article and book by

Jeff Pollard: J.W. Pollard 'Is Weismann's Barrier Absolute?' In *Beyond neo-Darwinism: Introduction to the New Evolutionary Paradigm*. (Eds) M.W. Ho and P.T. Saunders, Academic Press, London, pp 291–315, 1984; J.W. Pollard (Ed.) *Evolutionary Theory: Paths into the Future*. John Wiley & Sons, Chichester, UK, 1984.

2 E.J. Steele (1979). *Somatic Selection and Adaptive Evolution: On the Inheritance of Acquired Characters*. Williams and Wallace, Toronto. 2nd edition University of Chicago Press, Chicago, 1981.

3 S.J. Gould. *Dinosaur in a Haystack*. Penguin, Ringwood, Vic, p. 294, 1997. See also S.J. Gould. *Life's Grandeur*. Vintage Books, London, pp 221–222, 1997.

4 See S.J. Gould. *Life's Grandeur*. Vintage Books, London, p 222, 1997.

5 He subsequently published an 'omnibus' collection of excerpts from many of his published writings in *Bricks to Babel*. Hutchinson, London, 1981.

6 Richard Dawkins, *The Extended Phenotype*. Oxford University Press, Oxford, pp. 164–165, 1982.

7 Stephen Jay Gould. *Dinosaur in a Haystack*. Penguin, Ringwood, Vic, p. 128, 1997.

8 Some idea of the intensity of this controversy can be gleaned by two published pieces, one a riposte and the other an interview bracketing events from 1981 to 1996: see T. Steele 'Lamarck and Immunity; a Conflict Resolved' *New Scientist* vol 90: 360–361, 1981; and E.J. Steele 'Lamarckism: Still Alive and Thriving' *AEON* vol IV(3): 39–49, 1996 (interview).

9 Reviewed in E.J. Steele, R. M. Gorczynski and J.W. Pollard 'The Somatic Selection of Acquired Characters' In: *Evolutionary Theory: Paths into the Future*. Ed. J.W. Pollard. John Wiley, London, pp 217–237, 1984; and H. Rothenfluh and T. Steele 'Lamarck,

Darwin and the Immune System' *Today's Life Science* vol 5(7): 8–15 and vol 5(8): 16–22, 1993. Indeed, classical experiments thoughout the 1960s, 1970s and 1980s involving various immunological phenomena arising from the selective breeding of so-called 'Biozzi' high and low antibody-responder mice are consistent with an interpretation whereby antigenic stimulation of parents is somehow influencing the 'immunogenetic' variability in the offspring (see E.J. Steele 'Idiotypes, allotypes and a paradox of inheritance' In: *Paradoxes in Immunology.* Eds G.W. Hoffman, J.G. Levy, and G.T. Nepom. CRC Press Inc., pp 243–252, 1986).

10 Some of the emotive atmosphere can be gleaned from E.J. Steele 'Lamarckism: Still Alive and Thriving' *AEON* vol IV(3): 39–49, 1996 (interview) and a news report in a July 1981 issue of *Science* (R. Lewin 'Lamarck will not lie down' *Science* vol 213: 316–321, 1981).

11 TAG, TAA or TGA, see Appendix.

12 At present mainstream thinking distinguishes the processes of somatic hypermutation in mouse and human B lymphocytes (resulting in 'somatic point mutations') from chicken 'gene conversion' (resulting in segmental sequence transfers). We consider there are good grounds for thinking that they are linked by reverse transcription and homologous recombination processes as outlined in the RT-mutatorsome concept (Figure 5.6); see R.V. Blanden and E.J. Steele 'A Unifying Hypothesis for the Molecular Mechanism of Somatic Mutation and Gene Conversion in Rearranged Immunoglobulin Variable Genes' *Immunology and Cell Biology*, vol 76: 288–293, 1998.

13 Recently in a communication published in the *Journal of Molecular Biology* (vol 256: 813–817) entitled 'The Imprint of Somatic Hypermutation on the Repertoire

of Human Germline V Genes', Ian Tomlinson, Greg
Winter and their colleagues compared patterns of
somatic hypermutation established in antibodies
raised in mice against small chemical haptens (see
Chapter 3) with patterns of variation in human
germline V genes. They have concluded that, since
these patterns are different but complementary with
respect to the location of amino acid changes in the
antigen-combining site of the immunoglobulin, these
data are incompatible with homologous recombination between somatically derived DNA and germline
V genes proposed by us. We wish to raise two points.
First, most work on somatic hypermutation has
involved synthetic haptens. Such molecules could not
have been involved in the evolution of germline IgV
genes. Until much more data is accumulated
on patterns of somatic hypermutation using protein
and carbohydrate epitopes that were involved in
the evolution of adaptive immunity, their conclusion
is premature. Indeed, it is now getting clearer
from the work of Rolf Zinkernagel and colleagues
on mouse antibody responses to viral protein antigens
that little affinity maturation, and thus somatic
hypermutation, appears to occur in the generation
of such antibodies, implying that the repertoire
of mouse germline V elements is already at optimum
'average affinity' when used in the context of V(D)J
rearrangements resulting in H+L antibody
heterodimers. Second, in their paper Tomlinson
et al. referenced only one of our papers (Rothenfluh
et al. *Proceedings National Academy of Science
USA* vol 91: 12163–12167, 1994) and their consideration only of patterns of mutation in somatically
derived and germline IgV sequences provides
a narrow perspective for consideration of the issue
of possible transmission of somatic sequences to

the germline. Indeed, a much more comprehensive analysis of Ig sequences with respect to the somato-germline idea for vertebrates is provided in Rothenfluh et al. 'Evolution of V genes: DNA Sequence Structure of Functional Germline Genes and Pseudogenes', which was published in the journal *Immunogenetics* in 1995. Finally, the haptens used by Karl Landsteiner (Chapter 3) and other immunologists are small epitopes and during the course of an immune response it makes sense that somatic mutation within the antibody-combining sites of heavy and light chain variable region genes could result in selection of new amino acids lining a binding 'pocket', that cause an increase in affinity. A more recent analysis published in *Science* by Gary Wedemayer and colleagues supports this view (vol 276: 1665–1669, 1997).

14 G.F. Weiller, H.S. Rothenfluh, P. Zylstra, L.M. Gay, H. Averdunk, E.J. Steele and R.V. Blanden 'Recombination Signature of Germline Immunoglobulin Variable Genes' *Immunology and Cell Biology* vol 76: 179–185, 1998.

15 See the classic book by S. Ohno. *Evolution by Gene Duplication*. Springer, Berlin, 1970.

16 If random mutation followed by selection of germline genes for function were the mechanism, it would favour single chain antibodies and no DNA rearrangement. Indeed, soma-to-germline feedback may have been needed for the evolutionary emergence of both H+L heterodimers and rearrangement of variable V(D)J genes.

17 N. Lonberg, L.D. Taylor and another sixteen coauthors 'Antigen-Specific Human Antibodies from Mice Comprising Four Distinct Genetic Modifications' *Nature* vol 368: 856–859, 1994.

CHAPTER 7 BEYOND THE IMMUNE SYSTEM?

1 F. Wood Jones. *Habit and Heritage*. Kegan Paul, Trench, Trubner & Co., London, p 16, 1943. Republished with the permission of the publisher, Kegan Paul International Ltd.

2 The phenomena involving transmission of chemically induced diabetes or acquired thyroid dysfunction can be found in K. Okamoto 'Apparent Transmittance of Factors to Offspring by Animals with Experimental Diabetes' In: *On the Nature and Treatment of Diabetes*. B.S. Leibel and G.A. Wrenshall (Eds). Excerpta Med. Amsterdam, pp 627–631, 1965; M.G. Goldner and G. Spergel 'On the Transmission of Alloxan Diabetes and other Diabetogenic Influences' *Advances in Metabolic Disorders* vol 6: 57–72, 1972; and J.L. Bakke, N.L. Lawrence, J. Bennett and S. Robinson 'Endocrine Syndromes Produced by Neonatal Hyperthyroidism, Hypothyroidism, or Altered Nutrition and Effects Seen in Untreated Progeny' In: *Perinatal Thyroid Physiology and Disease*. D.A. Fisher and G.N. Burrow (Eds). Raven Press, New York, pp 79–116, 1975. Also see the review of these transgenerational phenomena by J.H. Campbell 'Autoevolution' In: *Perspectives on Evolution*. R. Milkman (Ed.). Sinauer Associates, New York, pp 190–201, 1982.

3 One female mouse with a very high and sustained blood glucose level was found amongst 235 progeny out of normal females mated to Streptozocin-treated males (E.J. Steele 'Observations on Offspring of Mice Made Diabetic with Streptozocin' *Diabetes* vol 37: 1035–1043, 1988; see also the initial reports on the use of this spontaneously arising diabetic mouse in collaborative experiments with Kevin Lafferty's

group, S.J. Prowse, K.J. Lafferty, C.J. Simeonovic, M. Agostino, K.M. Bowen and E.J. Steele 'The Reversal of Diabetes by Pancreatic Islet Transplantation' *Diabetes* vol 31 (Suppl. 4): 30–37, 1982; S.J. Prowse, E.J. Steele and K.J. Lafferty 'Islet Allografting Without Immunosuppression. Reversal of Insulitis-associated and Spontaneous Diabetes in Nonimmuno-suppressed Mice by Islet Allografts' *Aust. J. Exp. Biol. Med. Sci.* (now *Immunology and Cell Biology*) vol 60: 619–627, 1982.

4 J.H. Campbell 'Autoevolution' In: *Perspectives on Evolution.* R. Milkman (Ed.). Sinauer Associates, New York, pp 190–201, 1982.

5 The skin texture of the 'pads' on the soles of the feet of newborn human infants might also be placed in this class.

6 For a recent news update on the 'directed mutation' phenomena in bacteria see the September 1997 issue of *Scientific American* ('Evolution evolving' by Tim Beardsley, pp 9–12).

7 E.J. Steele and J. Cairns 'Dispute Resolved'. *Nature* vol 340: 336, 1989. Earlier that year Howard Temin reviewed the evidence for the existence of reverse transcriptase enzymes in bacteria (see H.M. Temin 'Retrons in Bacteria' *Nature* vol 339: 254–255, 1989).

8 See O. Prem Das, S. Levi-Minzi, M. Koury, M. Benner and J. Messing 'A Somatic Gene Rearrangement Contributing to Genetic Diversity in Maize' *Proceedings National Academy of Science USA* vol 87: 7809–7813, 1990. See also the review by D.M. Lambert, P.M. Stevens, C.S. White, M.T. Gentle, N.R. Phillips, C.D. Millar, J.R. Barker and R.D. Newcomb 'Phenocopies, Hereditary and Evolution' *Evolutionary Theory* vol 8: 285–304, 1989.

9 J. Waugh, MSc. Thesis 'An investigation into the induction of heavy metal tolerance in three plant

species' Department of Zoology, University of Auckland, New Zealand, 1987.

10 C.A. Cullis 'Environmentally induced DNA changes' In: *Evolutionary Theory: Paths into the Future.* Ed. J.W. Pollard. John Wiley, London, pp 203–216, 1984.

11 B. McClintock 'Mechanisms that Rapidly Reorganize the Genome' *Stadler Symposium* vol 10: 25–48, 1978.

12 H.S. Rothenfluh 'Hypothesis: A Memory Lymphocyte-Specific Soma-to-Germline Genetic Feedback Loop' *Immunology and Cell Biology* vol 73: 174–180, 1995.

13 This may not be such a fanciful scenario because it is known that mouse sperm cells maturing in the epididymal canals of the testis have the 'spontaneous ability to take up exogenous DNA' which can be integrated into the genomic DNA of the sperm nucleus (G. Zoraqi and C. Spadafora 'Integration of Foreign DNA Sequences into Mouse Sperm Genome' *DNA and Cell Biology* vol 16: 291–300, 1997). The work by Spadafora and colleagues has been controversial in the past (since 1989) but the recent data look convincing and they are defining many of the steps of the DNA-uptake pathway.

14 The formal term for an amino acid replacement substitution is 'nonsynonymous' and the term for a silent base change not leading to an amino acid substitution is 'synonymous'.

15 W-H. Li and D. Graur. *Fundamentals of Molecular Evolution.* Sinauer Associates, Sunderland, MA, pp 69–70, 1991.

16 G.F. Weiller, H.S. Rothenfluh, P. Zylstra, L.M. Gay, H. Averdunk, E.J. Steele and R.V. Blanden 'Recombination Signature of Germline Immunoglobulin Variable Genes' *Immunology and Cell Biology.* vol 76: 179–185, 1998.

17 The general term 'retrosequence' is also often used

and is applied to any DNA sequence that shows evidence of having been through an RNA intermediate at some point in its evolutionary history.

18 W-H. Li and D. Graur. *Fundamentals of Molecular Evolution.* Sinauer Associates, Sunderland, MA, pp 187–188, 1991.

19 And of course a 'clean integration event' involving a donor cDNA where *no* mutations have arisen at all in the coding regions of the gene would leave no trace at all of its somatic cell origin.

20 Similar reasoning in principle can also be applied to the maintenance of multiple identical copies of the other types of common housekeeping genes involved in chromosome packaging (histone genes) and those specifying the ribosomes or the protein synthesing machines (the ribosomal RNA genes).

21 See T.G. Sekevich, J.J. Bugert, J.R. Sisler, E.V. Koonin, G. Darai and B. Moss 'Genome Sequence of a Human Tumorigenic Poxvirus: Prediction of Specific Host Response-evasion Genes' *Science* vol 273: 813–816, 1996.

22 See S. Bartl, D. Baltimore and I.L. Weissman 'Molecular Evolution of the Vertebrate Immune System' *Proceedings National Academy of Science USA* vol 91: 10769–10770, 1994.

23 This involves work in progress with our PhD scholar Ms Paula Zylstra.

EPILOGUE

1 See M. Kimura. *The Neutral Theory of Molecular Evolution.* Cambridge University Press, Cambridge, 1983. Structure–function correlates are *essential* in biological understanding, at the molecular, cellular and organismal levels. The use of the word 'neutral' is a contradiction in terms in the context of Darwin-

ian natural selection. Secondly, we have great difficulty with the presence of degenerate third position
'silent' changes within a codon (see Appendix) being
considered as evidence for random genetic drift of
mutant forms of a gene which are selectively neutral.
The very high rate of 'silent' base changes in
mammalian housekeeping genes (e.g. histone genes)
strongly indicates powerful natural selection conserving the function of the protein.

2 Charles Darwin himself, despite his theory of Pangenesis, contributed to anti-Lamarckian sentiments
by not publicly acknowledging his intellectual debt
to Lamarck.

3 See *The New York Review* 12 June 1997 pp 34–37,
and 26 June pp 47–52; and Dennett's riposte and
Gould's reply, 14 August 1997, pp 64–65.

4 D. Berlinski 'The Deniable Darwin' *Commentary* pp
19–29, June 1996.

5 Prompting Stephen Jay Gould to label Dennett as
Dawkins's 'lapdog', see *The New York Review* 12
June 1997 pp 34–37: 'If history, as often noted,
replays grandeurs as farces, and if T.H. Huxley truly
acted as "Darwin's bulldog", then it is hard to resist
thinking of Dennett, [in his book *Darwin's Dangerous
Idea*] as "Dawkins's lapdog".'

6 M. Denton. *Evolution: A Theory in Crisis*. Burnett
Books, London, p. 22, 1985.

7 S. Kauffman. *At Home in the Universe: The Search
for Laws of Complexity*. Penguin Books, Ringwood,
Vic, 1996; also see G. Johnson. *Fire in the Mind:
Science, Faith and the Search for Order*. Penguin
Books, Ringwood, Vic, 1997.

8 L. Margulis. *Symbiosis in Cell Evolution*. W.H. Freeman and Co., San Francisco, 1981.

9 The term 'genome tracking' is used extensively
by Rod Langman, see our review article E.J. Steele,

H.S. Rothenfluh, G.L. Ada and R.V. Blanden 'Affinity Maturation of Lymphocyte Receptors and Positive Selection of T Cells in the Thymus' *Immunological Reviews* vol 135: 5–49, 1993.

10 Often without regard for the wider ethical consequences. Indeed in a book just published by the noted biologist Mae-Wan Ho (of the Open University, UK) the case is made against extreme reductionism and the rampant if not unhealthy alliance between molecular and cell biologists and the genetic engineering/biotechnology industry (M.W. Ho. *Genetic Engineering: Dream or Nightmare?* Gateway Books, Bath, UK, 1997).

11 This idea, as far as the authors are aware, was first discussed with Jeff Pollard in Toronto, Canada, in November–December 1978.

12 They are non-infectious as they have been genetically 'crippled'—once they infect a cell they cannot then go on and infect other cells. The late Howard Temin and his colleagues were pioneers in the development of these gene vectors.

13 See the 'Postscript 1981' In E.J. Steele. *Somatic Selection and Adaptive Evolution.* University of Chicago Press, Chicago, 2nd edition, 1981.

14 See the last chapter in E.J. Steele. *Somatic Selection and Adaptive Evolution.* Williams and Wallace, Toronto, 1979.

BIBLIOGRAPHY

Ada, G.L. and Nossal, G. (1987) 'The Clonal Selection Theory' *Scientific American* vol 255: 50–57.

Alexander, R.D. (1979) *Darwinism and Human Affairs.* University of Washington Press, Seattle.

Anderson, M.K., Shamblott, M.J., Litman, R.T. and Litman, G.W. (1995) 'Generation of Immunoglobulin Light Chain Gene Diversity in *Raja erinacea* Is Not Associated with Somatic Rearrangement, an Exception to a Central Paradigm of B Cell Immunity' *Journal of Experimental Medicine* vol 182: 109–119.

Bachmann, M.F., Kalinke, U., Althage, A., Freer, G., Burkhart, C., Roost, H-P., Aguet, M., Hengartner, H. and Zinkernagel, R.M. (1997) 'The Role of Antibody Concentration and Avidity in Antiviral Protection' *Science* vol 276: 2024–2027.

Bakke, J.L., Lawrence, N.L., Bennett, J. and Robinson, S. (1975) 'Endocrine Syndromes Produced by Neonatal Hyperthyroidism, Hypothyroidism, or Altered Nutrition and Effects Seen in Untreated Progeny' In: *Perinatal Thyroid Physiology and Disease.* D.A. Fisher and G.N. Burrow (Eds). Raven Press, New York, pp 79–116.

Bartl, S., Baltimore, D. and Weissman, I.L. (1994) 'Molecular Evolution of the Vertebrate Immune System' *Proceedings National Academy of Science USA* vol 91: 10769–10770.

Beardsley, T. (1997) 'Evolution Evolving' *Scientific American* vol 277: 9–12.

Berlinski, D. (1996) 'The Deniable Darwin' *Commentary* June pp 19–29.

Blanden, R.V. and Steele, E.J. (1998) 'A Unifying Hypothesis for the Molecular Mechanism of Somatic Mutation and Gene Conversion in Rearranged Immunoglobulin Variable Genes' *Immunology and Cell Biology* vol 76: 288–293.

Cairns, J., Overbaugh, J. and Miller, S. (1988) 'The Origin of Mutants' *Nature* vol 335: 142–145.

Campbell, J.H. (1982) 'Autoevolution' In: *Perspectives on Evolution*. R. Milkman (Ed.). Sinauer Associates, New York, pp 190–201.

Cannon, H.G. (1959) *Lamarck and Modern Genetics*. Manchester University Press, Manchester.

Cech, T.R. (1986) 'RNA as an Enzyme' *Scientific American* vol 255: 76–84.

Chen, C., Nagy, Z., Prak, E.L. and Weigert, M. (1995) 'Immunoglobulin Heavy Chain Gene Replacement: A Mechanism of Receptor Editing' *Immunity* vol 3: 747–755.

Cohn, M. (1968) 'The Molecular Biology of Expectation' In: *Nucleic Acids in Immunology*. O.J. Plescia and W. Braun (Eds). Springer-Verlag, New York, pp 671–715.

——(1974) 'A Rationale for Ordering the Data on Antibody Diversity' In: *Progress in Immunology II: Biological Aspects* I.L. Brent and J. Holborow (Eds). North-Holland Pub. Co., Amsterdam, pp 261–284.

Collins Dictionary of Biology 2nd Edition (1995) Hale, W.G., Margham, J.P. and Saunders, V.A. (Eds). Harper Collins Publishers, Glasgow, UK.

Crick, F.H. and Orgel, L.E. (1973) 'Directed Panspermia' *Icarus* vol 19: 341–346.

Cullis, C.A. (1984) 'Environmentally Induced DNA Changes' In: *Evolutionary Theory: Paths into the Future*. J.W. Pollard (Ed.). John Wiley, London, pp 203–216.

Cunningham, A.J. (1977) 'Evolution in Microcosm: The Rapid Somatic Diversification of Lymphocytes' *Cold Spring Harbor Symposium on Quantitative Biology* vol 41: 761–770.

Darwin, C. (1859) *The Origin of Species*. New American Library, Mentor Edition, 1958.

Darwin, C. (1868) *The Variation of Animals and Plants Under Domestication.* 2 vols, John Murray, London.

Davis, B.D., Dulbecco, R., Eisen, H.N., Ginsberg, H.S. and Wood, W.B. Jr (1969) *Microbiology.* Harper and Row, New York.

Dawkins, R. (1982) *The Extended Phenotype.* Oxford University Press, Oxford.

——(1986) *The Blind Watchmaker.* Longman, London.

Dennett, D.C. (1995) *Darwin's Dangerous Idea.* Penguin Books, Ringwood, Vic.

Denton, M. (1985) *Evolution: A Theory in Crisis.* Burnett Books, London.

Detlefsen, J.A. (1925) 'The Inheritance of Acquired Characters' *Physiological Reviews* vol 5: 244–278.

Eigen, M., Gardiner, W., Schuster, P. and Winkler-Oswatitsch, R. (1981) 'The Origin of Genetic Information' *Scientific American* vol 244: 78–94.

Eldredge, N. and Gould, S.J. (1972) 'Punctuated equilibria: an alternative to phyletic gradualism' In: *Models in Paleobiology.* T.J.M. Schopf (Ed.). Freeman, Cooper and Co., San Francisco, pp 82–115.

Fenner, F., Henderson, D.A., Arita, I., Jezek, Z. and Ladnyi, I.D. (1988) *Smallpox and its Eradication.* World Health Organisation.

Foote, J. and Eisen, H.N. (1995) 'Kinetic and Affinity Limits on Antibodies Produced During Immune Responses' *Proceedings National Academy of Science USA* vol 92: 1254–1256.

Gesteland, R.F. and Atkins, J.F. (Eds) (1993) *The RNA World.* Cold Spring Harbor Laboratory Press, Cold Spring Harbor, New York.

Gillman, M.A. (1996) *Envy as a Retarding Force in Science.* Avebury, Aldershot, UK.

Goldner, M.G. and Spergel, G. (1972) 'On the Transmission of Alloxan Diabetes and Other Diabetogenic Influences' *Advances in Metabolic Disorders* vol 6: 57–72.

Goldschmidt, R. (1940) *The Material Basis of Evolution.* Yale University Press, New Haven.

Gould, S.J. (1991) *Wonderful Life: the Burgess Shale and the nature of history.* Penguin Books, Ringwood, Vic.

——(1997) *Dinosaur in a Haystack.* Penguin Books, Ringwood, Vic.

——(1997) *Life's Grandeur: the spread of excellence from Plato to Darwin.* Vintage Books, London.

Greenberg, A.S., Avila, D., Hughes, M., Hughes, A., McKinney, E.C. and Flajnik, M.F. (1995) 'A New Antigen Receptor Gene Family that Undergoes Rearrangement and Extensive Somatic Diversification in Sharks' *Nature* vol 374: 168–173.

Henderson's Dictionary of Biological Terms 11th Edition (1995) Lawrence, E. (Ed.). Addison Wesley Longman Ltd, Harlow, Essex, UK.

Hinds-Frey, K.R., Nishikata, H., Litman, R.T. and Litman, G.W. (1993) 'Somatic Variation Precedes Extensive Diversification of Germline Sequences and Combinatorial Joining in the Evolution of Immunoglobulin Heavy Chain Diversity' *Journal of Experimental Medicine* vol 178: 825–834.

Ho, M.W. (1997) *Genetic Engineering: Dream or Nightmare?* Gateway Books, Bath, UK.

Holland, H.D. (1997) 'Evidence for Life on Earth More Than 3850 Million Years Ago' *Science.* vol 275: 38–39.

Hopkins, D.R. (1983) *Princes and Peasants: Smallpox in History.* University of Chicago Press, Chicago.

Hoyle, F. and Wickramasinghe, N.C. (1978) *Life Cloud: The Origin of Life in the Universe.* Sphere Books, London.

——(1996) *Our Place in the Cosmos.* Phoenix Paperback, Orion Books Ltd, UK.

Jablonka, E. and Lamb, M.J. (1994) *Epigenetic Inheritance and Evolution: The Lamarckian Dimension.* Oxford University Press, Oxford.

Jerne, N.K. (1973) 'The Immune System' *Scientific American* vol 229: 52–60.

Johnson, G. (1997) *Fire in the Mind: Science, Faith and the Search for Order.* Penguin Books, Ringwood, Vic.

Judson, H.F. (1979) *The Eighth Day of Creation.* Simon and Schuster, New York.

Kauffman, S. (1996) *At Home in the Universe: The Search for Laws of Complexity*. Penguin Books, Ringwood, Vic.

Kephart, J.O., Sorkin, G.B., Chess, D.M. and White, S.R. (1997) 'Fighting computer viruses' *Scientific American* vol 277: 56–61.

Kimura, M. (1983) *The Neutral Theory of Molecular Evolution*. Cambridge University Press, Cambridge.

Kitcher, P. (1996) *The Lives to Come: The Genetic Revolution and Human Possibilities*. Penguin Books, Ringwood, Vic.

Klenerman, P., Hengartner, H. and Zinkernagel, R.M. (1997) 'A Non-retroviral RNA Virus Persists in DNA Form' *Nature* vol 390: 298–301.

Koestler, A. (1971) *The Case of the Midwife Toad*. Hutchinson, London.

——(1978) *Janus. A Summing Up*. Hutchinson, London.

——(1981) *Bricks to Babel*. Hutchinson, London.

Kuby, J. (1997) *Immunology* 3rd edition. W.H. Freeman & Co., New York.

Kuhn, T.S. (1977) *The Essential Tension. Selected Studies in Scientific Tradition and Change*. University of Chicago Press, Chicago.

Lamarck, J.B. (1809) *Zoological Philosophy*. Translated by Hugh Elliot. University of Chicago Press, Chicago, 1984.

Lambert, D.M., Stevens, P.M., White, C.S., Gentle, M.T., Phillips, N.R., Millar, C.D., Barker, J.R. and Newcomb, R.D. (1989) 'Phenocopies, Heredity and Evolution' *Evolutionary Theory* vol 8: 285–304.

Langman, R.E. and Cohn, M. (1987) 'The E-T (Elephant-Tadpole) Paradox Necessitates the Concept of a Unit of B-Cell Function: The Protecton' *Molecular Immunology* vol 24: 675–697.

Lewin, R. (1981) 'Lamarck Will Not Lie Down' *Science* vol 213: 316–321.

Li, W-H. and Graur, D. (1991) *Fundamentals of Molecular Evolution*. Sinauer Associates, Sunderland, MA.

Litman, G.W. (1996) 'Sharks and the Origin of Vertebrate Immunity' *Scientific American* vol 275: 47–51.

Lonberg, N., Taylor, L.D. et al. (another 16 co-authors) (1994) 'Antigen-Specific Human Antibodies from Mice

Comprising Four Distinct Genetic Modifications' *Nature* vol 368: 856–859.

Madigan, M.T. and Marrs, B.L. (1997) 'Extremophiles' *Scientific American* vol 276: 66–71.

Margulis, L. (1981) *Symbiosis in Cell Evolution*. W.H. Freeman and Co., San Francisco.

McClintock, B. (1978) 'Mechanisms that Rapidly Reorganize the Genome' *Stadler Symposium* vol 10: 25–48.

McGraw-Hill Dictionary of Bioscience (1997) S.P. Parker (Ed.). McGraw-Hill, New York.

Medvedev, Z.A. (1969) *The Rise and Fall of T.D. Lysenko*. Translated by I.M. Lerner. Columbia University Press, New York.

Mehra, J. (1994) *The Beat of a Different Drum: The Life and Science of Richard Feynman*. Clarendon Press, Oxford.

Milburn, G. (1998) *The Feynman Processor: An Introduction to Quantum Computation*. Allen & Unwin, Sydney.

Mullis, K.B., Ferre, F. and Gibbs, R.A. (Eds). (1994) *The Polymerase Chain Reaction*. Birkhauser, Boston.

Neuberger, M.S. and Milstein, C. (1995) 'Somatic Hypermutation' *Current Opinion in Immunology* vol 7: 248–254.

Ohno, S. (1970) *Evolution by Gene Duplication*. Springer, Berlin.

Okamoto, K. (1965) 'Apparent Transmittance of Factors to Offspring by Animals with Experimental Diabetes' In: *On the Nature and Treatment of Diabetes*. B.S. Leibel and G.A. Wrenshall (Eds). Excerpta Med., Amsterdam, pp 627–631.

Oxford Dictionary of Biology 3rd Edition (1996) Martin, E., Ruse, M. and Holmes, E. (Eds). Oxford University Press, Oxford and New York.

Parham, P. (1995) 'A Boost to Immunity from Nurse Sharks' *Current Biology* vol 5: 696–699.

Pasteur, L. (1881) 'Vaccination in Relation to Chicken-cholera and Splenic Fever' *Transactions of the International Medical Congress*, 7th session, London, Aug 2–9 vol 1: 85–90.

Pauling, L. (1940) 'A Theory of the Structure and Process of Formation of Antibodies' *Journal American Chemical Society* vol 62: 2643–2657.

Penguin Dictionary of Biology. 9th Edition (1994) M. Thain and M. Hickman (Eds). Penguin Books, Ringwood, Vic.

Pollard, J.W. (1984) 'Is Weismann's Barrier Absolute?' In: *Beyond neo-Darwinism: Introduction to the New Evolutionary Paradigm.* M.W. Ho and P.T. Saunders (Eds). Academic Press, London, pp 291–315.

Pollard, J.W. (Ed.) (1984) *Evolutionary Theory: Paths into the Future.* John Wiley & Sons, Chichester, UK.

Prem Das, O., Levi-Minzi, S., Koury, M., Benner, M. and Messing, J. (1990) 'A Somatic Gene Rearrangement Contributing to Genetic Diversity in Maize' *Proceedings National Academy of Science USA* vol 87: 7809–7813.

Prowse, S.J., Lafferty, K.J., Simeonovic, C.J., Agostino, M., Bowen, K.M. and Steele, E.J. (1982) 'The Reversal of Diabetes by Pancreatic Islet Transplantation' *Diabetes* vol 31 (Suppl. 4): 30–37.

Prowse, S.J., Steele, E.J. and Lafferty, K.J. (1982) 'Islet Allografting Without Immuno-suppression. Reversal of Insulitis-associated and Spontaneous Diabetes in Non-immuno-suppressed Mice by Islet Allografts' *Aust. J. Exp. Biol. Med. Sci.* (now *Immunology and Cell Biology*) vol 60: 619–627.

Radman, M. and Wagner, R. (1988) 'The High Fidelity of DNA Replication' *Scientific American* vol 259: 24–30.

Reanney, D.C. (1986) 'Genetic Error and Genome Design' *Trends in Genetics* vol 2: 41–46.

Richardson, G. (1997) *Love as Conscious Action: Towards the New Society.* Gavemer Publishing, Sydney.

Roost, H-P., Bachmann, M.F., Haag, A., Kalinke, U., Pliska, V., Hengartner, H. and Zinkernagel, R.M. (1995) 'Early High-Affinity Neutralizing Anti-viral IgG Responses Without Further Improvements of Affinity' *Proceedings National Academy of Science USA* vol 92: 1257–1261.

Rothenfluh, H.S. (1995) 'Hypothesis: A Memory Lymphocyte-Specific Soma-to-Germline Genetic Feedback Loop' *Immunology and Cell Biology* vol 73: 174–180.

Rothenfluh, H.S., Blanden, R.V. and Steele, E.J. (1995) 'Evolution of V Genes: DNA Sequence Structure of Functional

Germ-line Genes and Pseudogenes' *Immunogenetics* vol 42: 159–171.

Rothenfluh, H.S. and Steele, E.J. (1993) 'Origin and Maintenance of Germ-line V-genes' *Immunology and Cell Biology* vol 71: 227–232.

Rothenfluh, H. and Steele, T. (1993) 'Lamarck, Darwin and the Immune System' *Today's Life Science* vol 5(7): 8–15 and vol 5(8): 16–22.

Sekevich, T.G., Bugert, J.J., Sisler, J.R., Koonin, E.V., Darai, G. and Moss, B. (1996) 'Genome Sequence of a Human Tumorigenic Poxvirus: Prediction of Specific Host Response-Evasion Genes' *Science* vol 273: 813–816.

Spergel, G., Khan, F. and Goldner, M.G. (1975) 'Emergence of Overt Diabetes in Offspring of Rats with Induced Latent Diabetes' *Metabolism* vol 24: 1311–1319.

Steele, E.J. (1979) *Somatic Selection and Adaptive Evolution: On the Inheritance of Acquired Characters*. Williams and Wallace, Toronto. 2nd edition University of Chicago Press, Chicago, 1981.

——(1986) 'Idiotypes, allotypes and a paradox of inheritance' In: *Paradoxes in Immunology*. G.W. Hoffman, J.G. Levy and G.T. Nepom (Eds). CRC Press Inc., pp 243–252.

——(1988) 'Observations on Offspring of Mice Made Diabetic with Streptozocin' *Diabetes* vol 37: 1035–1043.

——(Ed.) (1991) *Somatic Hypermutation in V-Regions*. CRC Press, Boca Raton, FL.

——(1992) 'Alistair Cunningham and the generation of antibody diversity *after* antigen' *Immunology and Cell Biology* vol 70: 111–117.

——(1996) 'Lamarckism: Still Alive and Thriving (Interview)' *AEON* vol IV(3): 39–49.

Steele, E., Blanden, R. and Rothenfluh, H. (1996) 'How Have Antibody Genes Evolved?' *Australasian Science* vol 17(4): 46–49.

Steele, E.J. and Cairns, J. (1989) 'Dispute resolved' *Nature* vol 340: 336.

Steele, E.J., Gorczynski, R.M. and Pollard, J.W. (1984) 'The Somatic Selection of Acquired Characters' In: *Evolutionary*

269

Theory: Paths into the Future. J.W. Pollard (Ed.). John Wiley, London, pp 217–237.

Steele, E.J. and Pollard, J.W. (1987) 'Hypothesis: Somatic Hypermutation by Gene Conversion via the Error Prone DNA→RNA→DNA Information Loop' *Molecular Immunology* vol 24: 667–673.

Steele, E.J., Rothenfluh, H.S., Ada, G.L. and Blanden, R.V. (1993) 'Affinity Maturation of Lymphocyte Receptors and Positive Selection of T Cells in the Thymus' *Immunological Reviews* vol 135: 5–49.

Steele, E.J., Rothenfluh, H.S. and Blanden, R.V. (1997) 'Mechanism of Antigen-Driven Somatic Hypermutation of Rearranged Immunoglobulin V(D)J Genes in the Mouse' *Immunology and Cell Biology* vol 75: 82–95.

Steele, T. (1981) 'Lamarck and Immunity; a Conflict Resolved' *New Scientist* vol 90: 360–361.

Storb, U., Ritchie, K.A., O'Brien, R.L., Arp, B. and Brinster, R. (1986) 'Expression, Allelic Exclusion and Somatic Mutation of Mouse Immunoglobulin Kappa genes' *Immunological Reviews* vol 89: 85–101.

Temin, H.M. (1989) 'Retrons in Bacteria' *Nature* vol 339: 254–255.

Tomlinson, I.M., Walter, G., Jones, P.T., Dear, P.H., Sonnhammer, E.L.L. and Winter, G. (1996) 'The Imprint of Somatic Hypermutation on the Repertoire of Human Germline V Genes' *Journal of Molecular Biology* vol 256: 813–817.

Tonegawa, S. (1983) 'Somatic Generation of Antibody Diversity' *Nature* vol 302: 575–581.

Waugh, J. (1987) 'An investigation into the induction of heavy metal tolerance in three plant species' MSc thesis, Department of Zoology, University of Auckland, New Zealand.

Wedemayer, G.J., Pattern, P.A., Wang, L.H., Schultze, P.G. and Stevens, R.C. (1997) 'Structural Insights into the Evolution of an Antibody Combining Site' *Science* vol 276: 1665–1669.

Weiller, G.F., Rothenfluh, H.S., Zylstra, P., Gay, L.M., Averdunk, H., Steele, E.J. and Blanden, R.V. (1998) 'Recombination Signature of Germline Immunoglobulin

270

Variable Genes' *Immunology and Cell Biology.* vol 76: 179–185.

Wood Jones, F. (1943) *Habit and Heritage.* Kegan Paul, Trench, Trubner & Co., London.

——(1953) *Trends of Life.* Edward Arnold & Co., London.

Woese, C.R. (1994) 'Universal Phylogenetic Tree in Rooted Form' *Microbiological Reviews* vol 58: 1–9.

Zoraqi, G. and Spadafora, C. (1997) 'Integration of Foreign DNA Sequences into Mouse Sperm Genome' *DNA and Cell Biology* vol 16: 291–300.

INDEX

ABO blood groups, 227
acquired callousing, 190–2
acquired characters: genetic transmission of, 189; not inherited, 164
acquired immune deficiency syndromes, 88
acquired immunological (neonatal) tolerance, 169, 171
acquired inheritance, 60, 169–72, 189, 190–6, 208, 244; adaptive mutations in bacteria, 193, 196; and callousing, 190–2; and heavy metal tolerance in plants, 196; and squatting habit, 192–5; in plants, 196; paternal transmission experiments, 169–72
acquired somatic learning, 105–6
Ada, G. L. (Gordon), xvi, 99–100, 143, 248, 261
adaptive fine-tuning, 216
adaptive radiation, 214
adaptors, in translation (protein synthesis), 225
affinity maturation of antibodies, 99–100, 127, 144–7, 227; defined, 127, 227; in Germinal Centre, 145–7; mechanism of, 144
affinity, of antibody binding, 82
African warthog, 191

Agent Orange, 137
agglutination, 68
Agostino, M., 257
AIDS, 23, 31
Alexander, R. D., 5
allele, 127, 136, 227
allergies, 80
Althage, A., 249
Altman, Sydney, 27, 52, 198
Alvarez, Luis, 183
Alvarez, Walter, 183
amino acids, definition of, 17, *see also* Appendix, 227
Anderson, M. K., 247
antibodies, 79, 100, 105; as cell surface antigen-recognition molecules, 105; complex genetics of, 100; in protection, 79
antibody diversity, 77–8, 87, 101, 115, 117–18; and potential repertoire size, 115, 117, 118; and repertoire problem/space in DNA, 77–8, 87; germline theory of, 101; somatic theory of, 101
antibody genes, and conventional genetics, 75–7
antibody response, time course of, 86
antibody specificity, 71, 102; and biological requirement for, 102; and Darwinian evolutionary

272

explanations, 71; and self
tolerance, 102
antibody, 10, 17, 61, 65–6, 69,
75–6, 97–8, 227; as a receptor
on B cell surface, 10, 97–8;
basis of specificity, 71;
definition of, 17, 61, 227; in
protection, 65; repertoire size,
69; structure, 66, 75–6
anticipatory purpose, concept of,
221
antigen, 59, 61, 228; definition
of, 61, 228; 'foreign', 59;
'self', 59
antigen-antibody complexes,
selective role in Germinal
Centres, 83, 145–7
antigen-binding site, *see also*
antigen-combining site:
definition of, 99, 228; and
variable (V) regions, 99; and
Ig, TCR variable regions, 139
antigen-combining site, 10, 62,
243
antigen-generated diversity,
controversial nature of, 148–9
antigen suicide experiment, and
clonal selection, 100
apoptosis, 127, 145–7, 228; and
Germinal Centre function,
145–7; defined, 127, 228
Aral Sea, pollution of, 136
archaea cells, and evolution, 20
Arguet, M., 249
Arita, I., 246
artificer of design, 220
Asian peoples, squatting habit,
192–5
asteroid impact craters, 182–3
asthma, 80
athymic children, 80
attenuation, of live vaccines, 64
Australian Aborigines, squatting
habit, 193–5
Australian National University,
181
autoimmune disease, 88, 218
Averdunk, H., 255, 258

Avery, Oswald, 245
Avila, D., 249

B lymphocyte (cell), 10, 59, 61,
79, 90–1, 98–9, 115, 217,
228; affinity-based competition
for antigen, 99; definition of,
61, 228; exponential increase
in cells clonally selected by
antigen, 98; 'one cell—one
antibody' concept, 115;
production in bone marrow,
90–1; in reproductive tissues,
217
Bachmann, M. F., 248–9
bacterial cells, 20, 32
Bakke, J. L., 190, 256
Baltimore, David, 11, 27, 36,
149, 168–9, 259
Barker, J. R., 257
Bartl, Simona, 168, 259
base (nucleotide), 229
base deletions, origin of stop
codons, 50–1
base pairing rule: and hereditary
mechanism, 37–40, 229; and
Chargaff's rules, 39
Basle Institute for Immunology,
149
Bateson, William, 14, 19
Beardsley, Tim, 257
Benner, M., 257
Bennett, J. C., 27, 110–11, 115
Bennett, J., 256
Berek, Claudia, 144
Berlinski, David, xvi, 210, 220,
260
Billingham, Rupert, 104–6
binding power, or avidity
(affinity) of antibody, 79, 82
Biozzi high-low responder mice,
253
Blackburn, Elizabeth, 56
Blanden, R. V. (Bob), 22, 27,
80, 100, 139, 157, 177,
247–51, 253, 255, 258, 261
blood groups (ABO), 67–8
blood transfusions, 67–8

blueprint, genetic or hereditary, and DNA,16, 28, 37, 41, 52
bone marrow, site production white blood cells, 89–91
booster inoculations, 86
Both, Gerry, xvi, 154
Bothwell, Al, 149
Bowen, K. M., 257
Breini, Friedrich, 92
Brenner, Sydney, 26, 41, 149
Brent, Leslie, 104–6
Bretscher, Peter, xvi, 27, 104
Bugert, J. J., 259
Burkhart, C., 249
Burnet, Frank Macfarlane, 26, 92, 94, 95, 99–101, 104–5, 150, 212
Butler, Samuel, 163, 192
Byrt, Pauline, 99–100

Cairns, John, 193, 196, 257
Cambrian explosion, 184, 214
Cambridge University, 129, 149
Campbell, J. H. (John), 190, 256–7
cancer, 137–8, 218
Cannon, H. G., 210
cap site, and initiation of transcription, 229
carbohydrates, 245
catalysts, 34–37, 229; as protein enzymes, 34–7; as RNA 'enzymes', 34–7; definition of, 34, 37, 229
cDNA, 179, 229, see also retrotranscript
CDR, 140–1, 176–7, see also antigen-binding site
Cech, Thomas, 27, 52, 198, 245
cell division, and origin mutations, 161
cell wall (in bacteria), 32
cells: energy generation within, 30, 34; growth and division, 30
cells, higher (eukaryotic), 20
cells, multicellular life and evolution, 20
Central Dogma, definition of,

xviii–xx, 23, 41–4, 46, 93, 206–7
central nervous system, compared to immune system, 89–90
chaperones, 92, 226
Chargaff, Erwin, 39
Chargaff's rules, 39
Chase, M., 26
Chen, C., 247
Chess, D. M., 248
chicken germline V pseudogenes, 120, 174–7; gene conversion program, 175; paradoxical structure of, 174–7; relation to neo-Darwinian paradigm, 177; Wu-Kabat structures in, 175–7
cholera, 67
chromosomes, 4; mixing paternal and maternal, 28, 32, 229
classical genetics, and somatic mutation, 126
clean integration event, 203, 259
Clonal Selection Theory, 10, 61, 94, 95–124, 102, 117, 212, 229
clone, a gene (e.g. insulin), 46
Cochrane, Ev, xvi
coding region, 230
codon, definition of, 17, 152, see also Appendix
Cohn, Melvin, xvi, 27, 75, 96, 101, 104, 118, 148, 239, 246, 247, 250
complementarity-determining region (CDR), 176–7, 230
computer analogies, 18, 28, 30–1, 34–6, 40–1, 45, 55, 121–4, 220–1
congenital abnormalities, 136
conservative base change, 221
Constant (C-) regions of immunoglobulins, 61, 151, 230
convergence, 215–16
Copernican revolution, 213
cowpox, 63
creationism, 15, 213
Crick, Francis, 26, 37, 39, 41, 47, 216, 245
Cullis, C. A., 258

Cunningham, Alistair, xvi, 27, 96, 101, 148–9, 250
Cuvier, George, 213
cytochrome c, and homologous genes, 18
cytochrome P-450, 217
cytokines, 61, 80, 231
cytoplasm, 32, 231
cytotoxic T cells, as killer T cells, 79

D-element, defined, 112–14
Darai, G., 259
Darwin, Charles, 1–5, 7–9, 15, 24, 101, 164, 187, 208, 211, 244 see also table on Dogma; and acquired inheritance, 2, 3, 8, 9; and concept of 'use and disuse', 7; Lamarckian thinking, 2, 3, 7, 9, 211; and origin of biological variation, 9, xviii–xx, see also table on Dogma; and origin of mutations, 132; and Pangenesis, 2, 3, 8, 9, 24, 164, 187, 211, see also table on Dogma; and pre-existing variation, 3, 101, see also table on Dogma; as British icon, 15; his central assumption, 3, see also table on Dogma; intellectual debt to Lamarck, 260; Origin of Species, 4, 5, 7
Darwin's bulldog (T. H. Huxley), 260
Darwinian fundamentalism, 210
Darwinian natural selection paradigm, 187
Darwinian revolution, success of, 209
Darwinian selection, 10, 23, 98, 145–8; and clonal selection idea, 98; in Germinal Centres, 145–8; in immune system, 10, 23
Darwinism: defined xviii–xx, see also table on Dogma; and

traditional explanations, 160, 171
Davis, B. D., 246
Dawkins' lapdog (Daniel C. Dennett), 260
Dawkins, Richard, 2, 167, 210–13, 220, 252
de Vries, Hugo, 19
definitions, see also Glossary, 227–43; of molecular and cellular terms, 17–18; of immunological terms, 17–18, 61–2, 127–8
Dennett, Daniel C., 2, 208, 210–11, 213, 260
Denton, Michael, 214, 260
Detlefsen, J. A., 244
devil's advocacy scenarios, xvi, 197
diabetes, 24; Alloxan-induced, 189, Streptozocin-induced, 189
dinosaurs, extinction, 183
diploid, 4, 116, 231
directed mutation, 161–2, 183, 196, 211, 257
directed panspermia, 45
divine intervention, 22, 181, 220
divine purpose, 220
DNA molecules, in chromosomes, 32
DNA polymerase, defined, 128, 133–4, 149, 231
DNA (deoxyribonucleic acid), 15–17, 28, 37–8, 42–4, 47–9, 113–14, 231; as genetic or hereditary blueprint, 16, 28, 37–8; definition of, 17; double helix structure, 37–8; molecules in chromosomes, 32, 231; rearrangement of Ig and TCR genes, 113–14; sequence evidence for fact of evolution, 15; synthesis or replication of, 42–4, 232; three dimensional structure, 47–9
DNA-based models, and cell division, 160–1

dogma, definition of, xviii–xx; of neo-Darwinism, 209
Doherty, Peter, 81
domain (Ig), 232
double helix, 37–8
Dreyer, W. J., 27, 110–11, 115
Dulbecco, R., 246

Edelman, Gerald, 75
editing functions, of polymerase enzymes, 133–4
Ehrlich, Paul, 67, 88, 95
Ei/MAR, 151–2, 155–7, see also locus-specific device
Eigen, Manfred, 55, 133, 245
Eisen, H. N., 246, 248
Eldredge, Niles, 214
endogenous RNA retroviruses, 33, 166, 179, 185–6, 198–9; as gene shuttles, 179, 198–9; as gene vectors, 185–6
endosymbiosis, 215
enzymes, and multi-molecular machines, 29, 34–5, 225, 232
epidemiological patterns, 219
epigenetic inheritance, and Lamarckian inheritance, 205–6
error correction, of DNA/RNA sequences, 133–4
error rates, polymerase enzymes, 133–5
error-prone copying, 53, 55, 133–5, 149, 157, 199
error-prone DNA repair, 137
evolution, and: conciousness, 218; evidence for in DNA sequences, 15; increasing complexity on earth, 20; mental realism, 219; soma-germline 'life cycle' of V genes, 180
exons, defined, 107–8, 232
extinctions, 214
extremophiles, 20

FDC, Follicular Dendritic Cells, 144–5, 232
Fenner, Frank, xvi, 104, 246

Feynman, Richard, 123, 248
final causes, 220
five prime (5′), 17, 232
Flajnik, M. F., 249
flanking region DNA sequences, 62, 237
flax plant, 196
Follicular Dendritic Cell (FDC), 83, 144–5, 232
Foote, J., 248
frame shift mutations, 174
framework region (FW), 232
Freer, G., 249
Freud, Sigmund, 165

Galileo Galilei, 213
Galileo spacecraft, 182
Gardiner, W., 245
Gay, L. M., 255, 258
Gearhart, Patricia, 149, 158
gemmules, 8, 9
gene conversion, 154, 233; in chicken V(D)J genes, 154
gene duplication, 184
gene shuttles, 166
gene therapy, 220
Genetic Code, 17, 41, 44, see also Appendix; ancient universality of, 45; and 'Directed Panspermia', 45; degeneracy of, 199; origin of, 45
genetic drift, 4, 174, 260; molecular predictions of, 174; randomness of, 260
genetic engineering, 216
genetic information, flow of, Central Dogma, 42–4
genetic inheritance, mechanism of, 37–40
genetic isolation, and origin of new species, 4
genetic recombination, definition of, 4, 233
genetic responsibility, 24, 221
genetic storage strategies, 118–19
genome, definition of, 29, 233
genome tracking, 216

Gentle, M. T., 257
germ cells, 3, 178, 233;
    definition of, 17, 233; genes
    of, 3
Germinal Centre, 127, 142–8,
    145–7, 233; and affinity
    maturation, 145–7; and
    somatic hypermutation, 145–7;
    defined, 127, 233
germline 'Vesuvian' model, 204;
    and origin retropseudogenes,
    204; as neo-Darwinism, 204;
    as neo-Weismannism, 204
germline configuration, 112–14,
    127, 172
germline DNA, 9, 136, 173,
    184; and antibody gene
    mutations, 9; mutations, 136;
    V gene repertoire, evolutionary
    selection pressures, 184; V
    genes, deficit of stop codons,
    173; V genes, number, 173
germline V-element, 113, 141–2,
    172; defined, 113; Wu-Kabat
    structures, 141; not subject to
    direct selection, 142, 172
Gillman, Mark, 14
Ginsberg, H. S., 246
Goldner, M. G., 189, 256
Goldschmidt, Richard, 215
Gorczynski, R. M. (Reg), xvi,
    169, 171, 252
Gorczynski–Steele experiment, 171
Gould, Stephen Jay, 164, 168,
    210, 214, 252, 260
Graur, D. 259
Gray, David, 143–4
Greenberg, A. S., 249

H chain, heavy chain of antibody
    or Ig, 75, 110, 151, 234
H+L heterodimer, 254
Haag, A., 248
haemoglobin, 109; mutants of,
    131–2
Hall, B., 26
haploid, 4, 116, 234
hapten, 61, 69, 234

hapten-protein conjugates, 69
Harvard University, 193
Haurowitz, Felix, 92
helper T cell, 234, 242
Henderson, D. A., 246
Hengartner, H., 245, 248–9
hereditary mechanism, 37–8
herpes virus, 31
Hershey, A., 26
heterodimer, H+L, 234, 254–5
heuristic device, in theory
    building, 167
high fidelity copying (loop), 53,
    55, 133, 149
higher cells (eukaryotes), 32
Hinds-Frey, K. R., 249
histone, proteins, 127, 134, 199;
    genes, 259
HIV, 11, 31, 51, 135; employs a
    'Trojan horse' strategy, 51;
    integration into genome, 51
Ho, Mae-Wan, 252, 261
Holland, H. D., 20
Holland, John, 133
homologous gene, 18, 19, 234;
    and evolution, 18, 19;
    definition of, 18
homologous recombination,
    155–7, 159, 179–80, 202, 235;
    and RT-model, 155–7, 159;
    defined, 235; in gene
    replacement, 202
homologous sequence, 235
Hood, Leroy, 149
'hopeful monster', 215
Hopkins, D. R., 246
horizontal transmission of genes,
    205, 215
housekeeping genes, 62, 78,
    107–8, 199–203, 235, 259–60;
    and antibody repertoire
    problem, 78; and natural
    selection, 260; definition of,
    62, 235; evolution of,
    199–203; highly conserved,
    199–200
Hoyle, Fred, 20–1, 45, 215
Hughes, A., 249

Hughes, M., 249
Human Genome Project, 40, 165; cultural inheritance, and Lamarckian evolution, 164
Huxley, T. H., 260
hybridomas, also monoclonal antibodies, 127, 149–50, 235; defined, 127, 235
hypervariable regions, of V genes, 139

Ig (immunoglobulin): definition of, 235; genes, pre-existing specificity before antigen, 101; genes, unusual features, gene rearrangements, 105
Ig class switch, IgM to IgG, 84, 86
Ig genes, pre-existing specificity before antigen, 101
Ig genes, unusual features, gene rearrangements, 105
IgG, 75, in secondary response, 86
IgM, IgD, IgG, IgA, IgE, 236
IgV (variable) genes, evolution of, 72–3, 236
immune deficiencies, 'agammaglobulinaemia', 80–1
immune memory, 73, 99
immune response, to unexpected, 72
immune system: anatomy of, 89–91; and integrity of body, 58; evolution of, 72–5
immunoglobulin (Ig), definition of and classes of, 61–2, 230, 236
influenza virus, 31, 56, 135
inheritance of acquired characteristics, 1, 3, 6, 21, see also table on Dogma; and Darwin's ideas, 3; articulation by Lamarck, 3, 21
inheritance of acquired somatic mutations, 56, 161–2, 166, 217
inheritance, definition of, 4
inherited bone facets, 192–5; in

Asian peoples, 192–5; in Australian Aborigines, 193–5; in Caucasian peoples, 193
insertion/deletion events: and mutations, 174; and recombination errors, 181
instructionist theory, of antibody formation, 213
integration footprint, of soma-to-germline events, 154, 179
intelligent gene manipulator, 22, 181
interferons, 61, 80, 231
introns, defined, 107–8, 237
invertebrates, in evolution, 20
isozymes, 227, 250

J-C intron, 113–14, 151
J-element, defined, 112–14
Jablonka, Eva, 206
Jacob, F., 26
Jenner, Edward, 63–4
Jerne, Nils, 26, 95
Jezek, Z., 246
John Curtin School of Medical Research, 80–1, 148, 197
Johnson, G., 260
Jones, Frederic Wood, xvi, 163–4, 187–8, 190–2, 194–5, 210, 256
Judson, Horace Freelance, 27
junk DNA, 62, 78, 237
Jupiter, (planet), 182
Jurassic Park, 16

Kabat, Elvin, 27, 139
Kalinke, U., 248–9
Kammerer, Paul, 13–14, 210
Kauffman, Stuart, 215, 260
Kelsoe, Garnett, 144
Kephart, J. O., 248
Khorana, H. Gobind, 26, 41, 42
killer T cell, 62, 65, 79–80, 237, 242
Kimura, M., 209, 259
Kitcher, Philip, 216
Klein, Jan, 167

Klenerman, P., 245
Koestler, Arthur, xvi, 1, 14, 163–4, 192, 210
Kohler, George, 149
Koonon, E. V., 259
Kornberg, Arthur, 26, 133
Koury, M., 257
Kuby, J., 140
Kuhn, Thomas, xix

L chain, 237
L-V intron, defined, 113–14
La Trobe University, 132
Ladnyi, I. D., 246
Lafferty, K. J., 256–7
Lamarck, Jean-Baptiste de Monet, Chevalier de, 1, 2, 5, 169, 211, 244; reception of ideas, 2; idea of species transformation, 5
Lamarckian: concepts/mechanisms and immune system, 22–3, 26; evolution in plants, 196; explanation, 187; gene feedback, 56, 105, 120–1, 196, 221; ideas, and controversy, 206; ideas, in scientific explanations, 206; inheritance, 221; inheritance and cultural change, 164; inheritance, and retrofection, 197; inheritance, ethical and philosophical dilemmas, 220; interpretation in biology, 165; interpretation, problems of, 192; 'scare', and Richard Dawkins, 167, 211; soma-to-germline feedback, 214, 216; thinking, controversial nature of, 210; thinking, denouement, 210
Lamarckism, 1, 6–7, 11, 14–15, 21, 211, see also table on Dogma; and controversy, 11, 21; and gene feedback loops, 6, 15, 21; and rapid genetic transformation, 6, 21; and soma-to-germline feedback, 21; and somatic genetic variability

in immune system, 6, 21; and the generation (origin) of genetic variability, 6; as heresy, 1, 7, 167; association with T. D. Lysenko, 14, 15
Lamb, Marion, 206
Lambert, D. M., 257
Landsteiner, Karl, 67–70, 92, 255
Langman, R. E. (Rod), xvi, 27, 104, 118, 239, 247, 260
Lawrence, N. L., 256
Levi-Minzi, S., 257
Lewin, R., 253
Li, W.-H., 259
lifestyle choice, genetic consequences, 218, 220
ligase, 180
light (L) antibody chain, 75, 110, 151
Linnaean Society, 208
lipids, 245
Litman, G. W., 246–7, 249
Litman, R. T., 247, 249
locus-specific device, 127, 151–2, 155–7, 217, 237; Ei/MAR, 217; in somatic hypermutation, 217; for somatic hypermutation, defined, 127, 237
Lonberg, N., 255
low fidelity copying loop, 135
lymph node swelling, 92
lymph nodes, 89–91
lymphatic vessels, vs blood vessels, 90–1
lymphocyte migration model, 198
lymphoid system, 90
Lysenko, T. D., 14, 15, 244

McCarty, Maclyn, 245
McClintock, Barbara, 196, 258
McKinney, E. C., 249
MacLennan, Ian, 143–4
MacLeod, Colin, 245
Madigan, M. T., 20
Maitland, Charles, 63
maize, 196

malaria, and sickle-cell anaemia,
131–2
Margulis, Lynn, 215, 260
Marrs, B. L., 20
measles, 80
Medawar, Peter, 26, 104–6, 167,
169, 171
Medvedev, Z. A., 244
Mehra, J, 248
meiosis, mixing paternal and
maternal chromosomes, 4, 19,
184
melanoma, 138
memory B lymphocytes (cells),
98, 117; lymphocytes, and
germ cells, 198; lymphocytes,
and reproductive tissues,
185–6; lymphocytes, as gene
shuttles, 185–6; responses, 86
Mendel, Gregor, 19, 21
mental realism, 219
Meselson, M., 26
messenger RNA, also mRNA,
32–3, 42–4, 120, 107–8, *see
also* Appendix
Messing, J., 257
Metchnikoff, Elie, 67
meteor impact craters, 182–3, 214
MHC (transplantation) antigen,
definition of, 62, 237, 242
Milburn, G., 248
Millar, C. D., 257
Milstein, Cesar, 129, 149–50,
155, 250
mitosis, 28
mobile B cells, as gene shuttles,
179–80
monoclonal antibodies, from
hybridomas, 149
monomeric IgM, 82
Montagu, Lady Mary Wortley,
60, 63
Moss, B., 259
mouse hepatitis virus, 189
Mullbacher, Arno, xvi, 197
Mullis, Kary B., 16, 27
multi-molecular machines,
enzymes as, 29, 46

multiculturalism, 221
multigene V families, 108, 113
multiple sclerosis, 80, 88
multipoint binding, avidity/affinity
of, 82, 85
mutagen, 127, 238
mutant proteins, 131–2
mutant sequence, 47; and
Darwinian selection, 47
mutation, 4, 40, 50–1, 54–5,
100, 130–7; of DNA
sequences, repair of, 54–5; of
Ig genes, after antigen, 100;
defined, 4, 130–1; and
Darwinian survival, 130; as
base deletions, 50–1;
deleterious nature of, 130–2;
inheritance in DNA, 50–1;
origin of, 130–7
mutatorsome, defined, 128

Nagy, E. L., 247
natural selection, 1, 3, 4; and
evolution, definition of, 4; and
the 'blind reaper', 3; as
dogma, 1, *see also* table on
Dogma; as driving force of
evolution, 1
Nature *vs* Nurture, in immune
system, 89
neo-Darwinian: extremism,
208–10, 220; paradigm, 219;
theory, and conventional
explanations, 171; assumptions,
60; standard model, 202
neo-Darwinian explanation: of
directed mutation in bacteria,
196; of inherited bone facets,
193; of inherited callosities, 191
neo-Darwinism, xviii–xx, 2, *see
also* table on Dogma; and
pre-existing genetic variability,
6; and purifying selection, 200;
and variability, 136; as dogma,
209, 212; as religion, 209;
conventional paradigm, 182;
definition of, xviii–xx, 2, 191
see also table on Dogma;

dogma and somatic mutation, 126; dominance of, 4–5

neo-Lamarckian: concepts, 148, 162, 179; feedback, 209; model, and origin retropseudogenes, 204; paradigm, 219; soma-to-germline feedback, 216

neo-Lamarckism, definition of, xviii–xx, see also table on Dogma, 2, 187, 212

neo-Weismann, standard model, 202

neo-Weismannism, 212

neo-Weismannists, defined, 212

Neuberger, Michael, 129, 250

neutral theory of molecular evolution, 209; criticism of, 259–60

Newcomb, R. D., 257

Nirenberg, Marshall, 26, 41, 42

Nishikata, H., 249

non-random distribution of mutations, 152

non-random DNA sequence pattern, 173

non-random mutation, 'play God scenarios', 125–6

Nossal, Gus, 143

nuclease, 181

nucleic acids, definition of, 17, 238

nucleotide bases, 'basic genetic letters' of DNA sequence, 16, 23, 29, 37–8, 42–4, 238

nucleus, 32–3, 44, 238

Ohno, Susumu, 255

Okamoto, K., 189, 256

olfactory genes, 217

Ontario Cancer Institute, 169

Open University, 261

open reading frames (ORF), 110, maintenance in germline V genes, 174

organelles, 29, 32, 44, 238

Orgel, Leslie, 45, 245

origin of life, molecular, 53

ostrich, 191

Oxford University, 210

pangenes, 8, 9, see also table on Dogma

Pangenesis, definition of, xviii–xx, see also table on Dogma, 24

panspermia, directed, 45

paradox, of Lamarckian gene feedback loop, 121

Parham, P., 249

Pasteur, Louis, 64, 246

paternal transmission experiments, controversy, 170–1, 188

pathogens, 58, 79

Pauling, Linus, 92–3

PCR, polymerase chain reaction, 16, 238

pentameric IgM, 75, 82, 85–6

peptide, 238

peripheral immune system, 90–1

phagocytes, 62, 65–6, 79, 83, 90–1, 238

phagocytosis, 62, 65–6, 79, 238

pharmaceutical industry, and antibody repertoire, 69

Phillips, N. R., 257

play God scenario, 160

Pliska, V., 248

point mutation, 50–1, 127, 130–1, 152, 239; defined, 130–1; replacement changes, 152, see also Appendix; silent changes, 152, see also Appendix; somatic, 151

poliomyelitis, 64

Pollard, J. W. (Jeff), xvi, 27, 153, 171, 251–2, 261

polymerase, 128, 239

polymorphism, 227, 250

Popper, Karl, 160, 167

Porter, Rodney, 75

Prak, E. L., 247

precipitation reactions, 70

Prem Das, O., 257

premature termination, 109–10; and 'out-of-frame' sequence, 109; and stop codons, 109

primary response, 86
primordial soup, 20
processed pseudogenes, without introns, 109
promoter region, 151–2
proof-reading functions, of polymerases 133–6
Protecton Theory, 62, 118, 239; and limits to antibody repertoire, 118, 239; definition of, 62, 239
protein, 17, 47–9, 92, 131–2, 239; folding (conformation), in mutants, 131–2; synthesis, *see* Appendix; three dimensional structure, 47–9; definition of, 17, 239; folding pattern, 92
Prowse, S. J., 257
pseudogenes (V), pristine condition of, 173
Punctuated Equilibrium, 168, 214

quantum mechanics, wave-particle duality, 102

Radman, M., 250
RAG proteins, 114, 174
Raison, Robert, 72
Rajewsky, Klaus, 144, 149
random genetic drift, 260
random gene rearrangements, V-D-J, 118
reading frame, of base triplets, 109
Reanney, Daryl, 132, 250
rearranged variable region gene, 128, 151; also termed V(D)J, 239
rearrangement of Ig genes, 110–15
recombination activating genes (RAG), 114, 174
recombination signature, germline V genes, 179
replacement base change, 152, 199–200, 255, 258
replicase, 128, 239
replication of DNA, 42–4

reproductive advantage, and natural selection, 5
response to unexpected, 73
retrofection, as Lamarckian inheritance, 197
retrogenes, 109
retrogenetics, 205
retropseudogenes (cDNA), 201–3; as integrated retrotranscripts, 202; as molecular fossils, 202–3; definition of, 201
retrosequence, 258
retrotranscript (cDNA), 11, 52, 155–7, 179, 199, 204–5, 210, 241; and V regions, 210; in viral genomes, 204–5
retroviral gene vectors, 216
retroviruses, 10–11, 31, 56, 216; as gene vectors, 216; endogenous RNA retroviruses, 10
Reverse Transcriptase (RT-) Model, 153–4, 158–60; novel predictions of, 158–60; of somatic hypermutation, 153–4, 156–7; predictive power and robustness, 159–60
reverse genetics, 221, 245
reverse transcriptase: definition of, xviii–xx, *see also* table on Dogma; 56, 128, 133–4, 166, 198–9, 240; -based gene feedback loop, 196
reverse transcription, 9, 10, 42–4, 47–52, 109, 120, 155–7, 179, 186, 197–8; and fused germline VJ or VDJ sequences, 120; and HIV, 33; and soma-to-germline feedback, 120; and somatic hypermutation, 211; and Somatic Selection Theory, 9,10; defined 240; generalised, definition of, xviii–xx, *see also* table on Dogma, 47–52, 198; reception of concept, 11
reverse transcripts, cDNA, 155–7

reverse translation, 245
Rh disease of the newborn, 68
rheumatoid arthritis, 88
ribosomal RNA, 248, *see also*
    Appendix, genes, 259
ribosomes, and translation mRNA
    into protein, 43–4, *see also*
    Appendix, 108, 155, 224–6,
    240
ribozymes, 35, 52, 56, 107, 198,
    225, 240
RNA (ribonucleic acid): and
    transcription, 42–4; as
    pre-mRNA, 108, 113–14;
    copied from DNA template,
    42–4; defined, 128, 240;
    information carrier and
    catalyst, 36; intermediate,
    defined, 17, 41, 128, 240,
    259; intermediates, error-prone
    copying of, 153; molecules,
    evolution of, 55; relationship
    to DNA, 42–4; replicase,
    133–4; splicing, also termed
    processing, 107–8, 113–14,
    120; structure of, 42–4; three
    dimensional structure, 47–9;
    tumour viruses, as retroviruses,
    49; 'World', 52–7
Roberts, R., 27, 107
Robinson, S., 256
Rockefeller University, 245
Roost, H.-P., 248–9
Rothenfluh, H. S. (Harry), xvi,
    27, 157, 172, 177, 185, 198,
    247–52, 254–5, 258, 261
RT-model, 153–8, *see also*
    Reverse Transcriptase Model
RT-mutatorsome, defined, 128,
    240

Salk Institute for Biological
    Studies, 149
Santa Fe Institute, 215
Saunders, P. T., 252
Schuster, P., 245
science, definition of, xiii–xiv
secondary response, 86

Sekevich, T. G., 259
selection window, concept, 218
selectively permeable membrane,
    29, 32
self antigen, 228
self-tolerance, 73, 87–9, 96–7,
    102–5; and Darwinian negative
    selection, 105; and deletion of
    forbidden clones, 97; induction
    of, 96; mechanism, 102–5; not
    genetically determined, 103–4;
    self *versus* non-self
    discrimination, in self
    tolerance, 102
Selsing, Eric, 158
sex determination, 4
Shamblott, M. J., 247
Sharp, P., 27, 107
Shoemaker-Levy 9, 182–3
sickle-cell anaemia, 131–2, 136
silent base change, 152, 199–200,
    203, 225, 258, 260; and
    functional conservation, 203;
    as mutations, 152, 258, 260
Simeonovic, C. J., 257
single copy genes, 78, 107–8,
    199–203; and antibody
    repertoire problem, 78; and
    related retropseudogenes, 201;
    evolution of, 199–203
Sisler, J. R., 259
smallpox, 60–7, elimination of by
    WHO, 64
soma-to-germline feedback loop,
    2, 9, 10, 26, 105, 114, 120,
    128, 163–86, 198, 200–1,
    203–4, 208–10, 214–19, 241,
    248, 255; and cDNA
    retrotranscripts, 201, 248; and
    evolution of vertebrates, 179;
    and 'genetic impact' events or
    impact footprint, 180, 183,
    200; and Lamarckian
    mechanisms, 26; and lifestyle
    choice, 218; and V gene
    evolutionary 'life cycle', 180;
    as genetic channel, 216;
    beyond immune system, 209;

defined, 128, 241;
environmentally sensitive,
215–16; epidemiological
patterns, 219; ethical
implications, 219; evolutionary
significance, 183–5; general
consequences, 219; general
theory of, 203; implications for
mankind, 208; integration
footprint defined, 114;
mechanism of, 177–9;
transmission process, 2, 9, 10,
214
Somatic Selection Theory, 8, 10,
154, 166–8, 180, 182, 211–12;
and controversy, 211–12;
criticism of, 167–8; organising
principle, 168
somatic cell, definition of, 17, 241
somatic cells, genes of, 3
somatic configuration, defined,
112–14, 128, 172, 241
somatic hypermutation, 125–62;
after antigen stimulation,
142–3, 150; and gene
conversion, 249; and Germinal
Centres, 142–8; and reverse
transcription, 211; beneficial
nature of, 139, 142; confined
to V(D)J genes, 115–16, 119,
130, 142, 144, 151–2; defined,
128; distribution in V(D)J,
151; evolutionary significance,
129; in B lymphocytes, 142;
linkage to soma-to-germline
process, 186; of antibody
genes, 2; rate in V(D)J genes,
138; RT mechanism, 253;
significance in extant
vertebrates, 129, 185; stop
signal of, 156–7, 160–1
somatic mutation, 125–62; and
cancer, 137–8; as fact, 149;
biological purpose of, 126;
confined to V(D)J, 115, 116,
119, 130, 142, 144, 151–2;
controversy, 126; definition of,
xviii–xx, see also table on

Dogma; in V genes, 78;
inheritance of, 217; rate and
cell turnover, 137
somatic RNA processing
signature, and Weiller
algorithm, 182
somatic signature, in germline V
genes, 172–83
somatic vs germline, 75–7,
101–2, 120, 149;
configurations, significance to
data interpretation, 120;
debate, 149; debate in
immunology, 75–7, 149;
debate, philosophical
significance, 101–2
Sorkin, G. B., 248
Spadafora, C., 258
speciation, 214–16
specificity of antibodies, definition
of, 61
Spencer, Herbert, 163
Spergel, G., 189, 256
sperm cells, DNA-uptake
pathway, 258
Spiegelman, S., 26
spleen, 90
spliceosome, 107–8, 155
spontaneous abortions, 136
sports (mutations), 132
squatting habit, and inherited
bone facets, 192–5
Stahl, F., 26
Stalin, Joseph, 14
Steele, E. J. (Ted), 8, 21–2, 27,
52, 100, 139, 153–4, 157,
163, 168–9, 171, 188–9, 196,
210–12, 247–53, 255–8, 260–1
Stevens, P. M., 257
stop codons, definition of, 109,
see also Appendix
Storb, Ursula, 143, 149

T cell receptor (TCR), 65–6,
80, 241; genes, evolution of,
72–3
T cell recognition, 81, 87;
binding of self MHC plus

foreign peptide, 87; of
virus-infected cells, 81, cf.
binding sites of antibodies, 81
T helper cells, 234
T killer cells, 237
T lymphocytes, 59, 62, 79,
90–1, 98, 241, *see also* T cells;
exponential increase in cells
clonally selected by antigen,
98; in protection against viral
diseases, 79; origin of 90–1
Talmage, David, 26, 96
Taylor, L. D., 255
TCR (T cell receptor), 102,
105; genes, unusual features,
gene rearrangements, 105; and
self tolerance, 102; requirement
for, 102
Teilhard de Chardin, 211
telomeres, 56
Temin, H. M. (Howard), 11,
26–7, 36, 133, 198, 206–7,
257, 261
template molecule, defined, 128,
242
'thin-edge-of-the-wedge'
argument, 197, 217
Thompson, Craig B., 174
three prime (3'), 17, 242
thymus, 89–91
Tognetti, Keith, xvi
Tomlinson, I., 254
Tonegawa, Susumu, 27, 76, 78,
111, 115, 149, 247;
demonstration of V(D)J
rearrangement, 115
tools of molecular biology, and
text manipulation in word
processor, 16, 18
transcription, mRNA synthesis,
32–3, 42–4, 151–2, 242
transfer RNA, tRNA, 248, *see
also* Appendix
transgenerational phenomena,
189–90
transgenic mice, 129, 143
translation, of mRNA into

protein, 43–4, 242, *see also*
Appendix
transplantation antigens (MHC),
62, 103, 242; and self
tolerance, 103; definition of,
62; natural diversity of, 103
transport across membrane, 32–3
triplet code, 42, 44, 92 *see also*
Appendix
triplet codons, 175
truth, definition of, xiv
Tufts University, 210

UCLA Medical School, 190
unequal crossing over, 184
University of Birmingham, 143

V gene, defined, 243
V gene replacement, 119
V to C separation, evolutionary
significance of, 152–3
V(D)J (rearranged variable region
gene), 78, 112–14, 128, 151,
239, 243, 249, 255; and DNA
rearrangement, 249; and
somatic mutation machinery,
114; DNA rearrangements,
151, 255; generic definition,
112–14, 128, 243
vaccines, 60–7; and attenuated
strains, 64; recombinant DNA
technology and, 64;
vaccination, 60–7
variable (V) region genes, 10,
23, 62, 75, 78, 99, 243;
amino acid sequences, 99; and
DNA rearrangements, V(D)J,
78; definition of, 62; of
immune system, 23; somatic
mutation of, 23
variolation, 60–7
VD fusions in germline, chicken
V pseudogenes, 142
VDJ fused sequences, in germline
DNA, 120
VDJ fusions in germline,
cartilaginous fish, 142
vertebrate immune system, 59,

73, 93–4, 168; evolution of, 168; features of, 73, 93–4
vertebrates, and immune responses, 58; in evolution, 20
'Vesuvian' mode of evolution, 202
Vietnam War, 137
viral infections, consequences, 65
viruses, 32
VJ fused sequences, in germline DNA, 120
VJ fusions in germline, cartilaginous fish, 142

Wagner, R., 250
Wallace, Alfred Russell, 208, 244
Wallace, Ann, xvi
Walter and Eliza Hall Institute of Medical Research, 100, 143
Watson, James, 26, 37, 39, 47, 49, 216
Waugh, J., 257
Wedemayer, G., 255
Weigert, Martin, 119, 247
Weill, Jean-Claude, 174
Weiller, G. F. (Georg), 27, 179, 181, 255
Weiller recombination algorithm, 181–2
Weiller recombination signal, 200
Weismann, August, 12
Weismannism, definition of, xviii–xx, see also table on Dogma
Weismann's Barrier, 163–5,

167–8, 211; and circumcision, 12; and neo-Darwinism, 4; as genetic chastity belt, 165; definition of, xvi, xviii–xx, 1, 12, 19, see also table on Dogma; direct penetration of, 23, 170, 179, 185, 187, 205, 219
Weismann's dogma, 212
Weissman, Irving, 168–9, 259
White, C. S., 257
White, S. R., 248
Wickramasinghe, Chandra, 20–1, 45, 215
Winkler-Oswatitsch, R., 245
Winter, G., 254
Woese, C. R., 20
Wood, J. B. jr, 246
Wu, T. T., 27, 139
'Wu-Kabat structures' 139–41, 152, 166, 173, 181; and antigen binding selection, 141; defined as variability plot, 139–41, 173; in unrearranged germline V-elements, 141

yellow fever, 64

Zinkernagel, R. M. (Rolf), 52, 81, 126, 128, 245, 248–9, 254
Zoraqi, G., 258
Zylstra, Paula, xvi, 27, 255, 258–9